9301939

B. Hofmann-Wellenhof, H. Lichtenegger,
and J. Collins

Global Positioning System

Theory and Practice

Second edition

Springer-Verlag Wien New York

Prof. Dr. Bernhard Hofmann-Wellenhof
Dr. Herbert Lichtenegger
Abteilung für Landesvermessung und Landinformation, Technische Universität Graz
Graz, Austria

Dr. James Collins
GPS Services, Inc.
Rockville, Maryland, U.S.A.

© 1992 and 1993 Springer-Verlag Wien
Printed in Austria by Adolf Holzhausens Nachfolger, A-1070 Wien
Printed on acid-free paper

Cover illustration courtesy of Rockwell International

With 35 Figures

ISBN 3-211-82477-4 Springer-Verlag Wien New York
ISBN 0-387-82477-4 Springer-Verlag Wien New York

ISBN 3-211-82364-6 1. Aufl. Springer-Verlag Wien New York
ISBN 0-387-82364-6 1 st ed. Springer-Verlag New York Wien

We dedicate this book to

Benjamin William Remondi

Foreword

This book is dedicated to Dr. Benjamin William Remondi for many reasons. The project of writing a Global Positioning System (GPS) book was conceived in April 1988 at a GPS meeting in Darmstadt. Dr. Remondi discussed with me the need for an additional GPS textbook and suggested a possible joint effort. In 1989, I was willing to commit myself to such a project. Unfortunately, the timing was less than ideal for Dr. Remondi. Therefore, I decided to start the project with other coauthors. Dr. Remondi agreed and indicated his willingness to be a reviewer.

I selected Dr. Herbert Lichtenegger, my colleague from the University of Technology at Graz, Austria, and Dr. James Collins from the United States.

In my opinion, the knowledge of the three authors should cover the wide spectrum of GPS. Dr. Lichtenegger is a geodesist with broad experience in both theory and practice. He has specialized his research to geodetic astronomy including orbital theory and geodynamical phenomena. Since 1986, Dr. Lichtenegger's main interest is dedicated to GPS. Dr. Collins retired from the U.S. National Geodetic Survey in 1980, where he was the Deputy Director. For the past ten years, he has been deeply involved in using GPS technology with an emphasis on surveying. Dr. Collins was the founder and president of Geo/Hydro Inc. My own background is theoretically oriented. My first chief, Prof. Dr. Peter Meissl, was an excellent theoretician; and my former chief, Prof. DDDr. Helmut Moritz, fortunately, still is.

It is appropriate here to say a word of thanks to Prof. DDDr. Helmut Moritz, whom I consider my mentor in science. He is – as is probably widely known – one of the world's leading geodesists and is currently president of the International Union for Geodesy and Geophysics (IUGG). In the fall of 1984, he told me I should go to the U.S.A. to learn about GPS. I certainly agreed, although I did not even know what GPS meant. On the same day, Helmut Moritz called Admiral Dr. John Bossler, at that time the Director of the National Geodetic Survey, and my first stay in the U.S. was arranged. Thank you, Helmut! I still remember the flight where I started to read the first articles on GPS. I found it interesting but I did not understand very much. Benjamin W. Remondi deserves the credit for providing my GPS instruction. He was a very patient and excellent teacher. I benefited enormously, and I certainly accepted his offer to return to the U.S.A. several times. Aside from the scientific aspect, our families have also become friends.

The selection of topics is certainly different from the original book conceived by Dr. Remondi. The primary selection criteria of the topics were:

relevancy, tutorial content, and the interest and expertise of the authors. The book is intended to be a text on GPS, recognizing the tremendous need for textual materials for professionals, teachers, and for students. The authors believe that it was not necessary to dwell on the latest technical advances. Instead, concepts and techniques are emphasized.

The book can be employed as a classroom text at the senior or graduate levels, depending on the level of specialization desired. It can be read, selectively, by professional surveyors, navigators, and many others who need to position with GPS.

May 1992   B. Hofmann-Wellenhof

Preface

The contents of the book are partitioned into 13 chapters, a section of references, and a very detailed index which should immediately help in finding certain topics of interest.

The first chapter is a historical review. It shows the origins of surveying and how global surveying techniques have been developed. In addition, a short history on the Global Positioning System (GPS) is given.

The second chapter is an overview of GPS. The system is explained by means of its three segments: the space segment, the control segment, and the user segment.

The third chapter deals with the reference systems, such as coordinate and time systems. The inertial and the terrestrial reference frames are explained in the section on coordinate systems, and the transformation between them is shown. The definition of different times is given in the section on time systems, together with appropriate conversion formulas.

The fourth chapter is dedicated to satellite orbits. This chapter specifically describes GPS orbits and covers the determination of the Keplerian and the perturbed orbit, as well as the dissemination of the orbital data.

The fifth chapter covers the satellite signal. It shows the fundamentals of the signal structure with its various components and the principles of the signal processing.

The sixth chapter deals with the observables. The data acquisition comprises code and phase pseudoranges and Doppler data. The chapter also contains the data combinations, both the phase combinations and the phase/code range combinations. Influences affecting the observables are described. Examples are: the atmospheric and relativistic effects, multipath, and the impact of the antenna phase center.

The seventh chapter is dedicated to surveying with GPS. This chapter defines the terminology used and describes the planning of a GPS survey, surveying procedures, and in situ data processing.

The eighth chapter covers mathematical models for positioning. Models for observed data are investigated. Therefore, models for point positioning and relative positioning, based on various data sets, are derived.

The ninth chapter comprises the data processing and deals with the sophisticated cycle slip detection and repair technique. This chapter also includes the resolving of phase ambiguities. The method of least squares adjustment is assumed to be known to the reader and, therefore, only a brief review is included. Consequently, no details are given apart from the linearization of the mathematical models, which are the input for the adjustment procedure.

The tenth chapter links the GPS results to terrestrial data. The necessary transformations are given where the dimension of the space and the transformations are considered.

The eleventh chapter treats software modules. The intent of this chapter is not to give a detailed description of existing software and how it works. This chapter should help the reader decide which software would best suit his purposes. The very short sections of this chapter try to cover the variety of features which could be relevant to the software.

The twelfth chapter describes some applications of GPS. Global, regional, and local uses are mentioned, as well as the installation of control networks. The compatibility of GPS with other systems, such as Inertial Navigation Systems (INS) and the Global Navigation Satellite System (GLONASS), the Russian equivalent to GPS, is shown.

The thirteenth chapter deals with the future of GPS. Both critical aspects, such as selective availability and anti-spoofing, are discussed, along with positive aspects such as the combination of GPS with GLONASS and the International Maritime Satellite Communication Organization (INMARSAT). Also, some possible improvements in the hardware and software technology are suggested.

The hyphenation is based on Webster's Dictionary. Therefore, some deviations may appear for the reader accustomed to another hyphenation system. For example, the word "measurement", following Webster's Dictionary, is hyphenated mea-sure-ment; whereas, following The American Heritage Dictionary, the hyphenation is meas-ure-ment. The Webster's hyphenation system also contains hyphenations which are sometimes unusual for words with a foreign language origin. An example is the word "parameter". Following Webster's Dictionary, the hyphenation is pa-ram-e-ter. The word has a Greek origin, and one would expect the hyphenation pa-ra-me-ter.

Symbols representing a vector or a matrix are underlined. The inner product of two vectors is indicated by a dot "·". The outer product, cross product, or vector product is indicated by the symbol "×". The norm of a vector, i.e., its length, is indicated by two double-bars "‖".

Many persons deserve credit and thanks. Dr. Benjamin W. Remondi of the National Geodetic Survey at Rockville, Maryland, was a reviewer of the book. He has critically read and corrected the full volume. His many suggestions and improvements, critical remarks and proposals are gratefully acknowledged.

A second technical proofreading was performed by Dipl.-Ing. Gerhard Kienast from the section of Surveying and Landinformation of the University of Technology at Graz. He has helped us with constructive critique and valuable suggestions.

Nadine Collins kindly read and edited the book in its final form, improving the flow and grammar of the text.

The index of the book was produced using a computer program written by Dr. Walter Klostius from the section of Surveying and Landinformation of the University of Technology at Graz. Also, his program helped in the detection of spelling errors.

The book is compiled based on the text system LATEX. Some of the figures included were also developed with LATEX. The remaining figures are drawn by using Autocad 11.0. The section of Physical Geodesy of the Institute of Theoretical Geodesy of the University of Technology at Graz deserves the thanks for these figures. Dr. Norbert Kühtreiber has drawn one of these figures, and the others were carefully developed by Dr. Konrad Rautz. This shows that theoreticians are also well-suited for practical tasks.

We are also grateful to the Springer Publishing Company for their advice and cooperation.

Finally, the inclusion by name of a commercial company or product does not constitute an endorsement by the authors. In principle, such inclusions were avoided whenever possible. Only those names which played a fundamental role in receiver and processing development are included for historical purposes.

May 1992   B. Hofmann-Wellenhof   H. Lichtenegger   J. Collins

Preface to the second edition

The first edition was released in May 1992. Since then, the first and second printing have been completely sold. There was not sufficient time to prepare a revised version for the second edition. Therefore, only a few misspellings and errors were corrected and other minor improvements performed. However, the authors would appreciate your comments for consideration in the case of a fully revised and updated version.

March 1993   B. Hofmann-Wellenhof   H. Lichtenegger   J. Collins

Contents

Abbreviations

AC	Alternating Current
ACS	Active Control System
AFB	Air Force Base
AGREF	Austrian GPS Reference (network)
AOC	Auxiliary Output Chip
A-S	Anti-Spoofing
AVL	Automatic Vehicle Location
BC	Ballistic Camera
BDT	Barycentric Dynamic Time
C/A	Coarse Acquisition
CAD	Computer Aided Design
CEP	Celestial Ephemeris Pole
CIGNET	Cooperative International GPS Network
CIO	Conventional International Origin
CIS	Conventional Inertial System
CSOC	Consolidated Space Operations Center
CTS	Conventional Terrestrial System
DC	Direct Current
DD	Double-Difference
DEC	Digital Equipment Corporation
DGPS	Differential GPS
DMA	Defense Mapping Agency
DoD	Department of Defense
DOP	Dilution of Precision
ECEF	Earth-Centered-Earth-Fixed
FAA	Federal Aviation Administration
FGCC	Federal Geodetic Control Committee
FM	Frequency Modulated
GDOP	Geometric Dilution of Precision
GIS	Geographic Information System
GLONASS	Global Navigation Satellite System
GOTEX	Global Orbit Tracking Experiment
GPS	Global Positioning System
GPST	GPS Time
GRS	Geodetic Reference System
HDOP	Horizontal Dilution of Precision

HIRAN	High Range Navigation
HOW	Hand Over Word
IAG	International Association of Geodesy
IAT	International Atomic Time
IBM	International Business Machines (corporation)
IERS	International Earth Rotation Service
IGS	International GPS Geodynamics Service
INMARSAT	International Maritime Satellite (organization)
INS	Inertial Navigation System
ISU	International System of Units
IUGG	International Union for Geodesy and Geophysics
JD	Julian Date
JPL	Jet Propulsion Laboratory
JPO	Joint Program Office
LAN	Local Area Network
MCS	Master Control Station
MIT	Massachusetts Institute of Technology
MITES	Miniature Interferometer Terminals for Earth Survey
MJD	Modified Julian Date
MMIC	Monolithic Microwave Integrated Circuit
NAD	North American Datum
NASA	National Aeronautics and Space Administration
NAVSTAR	Navigation System with Time and Ranging
NGS	National Geodetic Survey
NNSS	Navy Navigational Satellite System
NSWC	Naval Surface Warfare Center
OCS	Operational Control System
OEM	Original Equipment Manufacturer
OTF	On-the-Fly
P	Precision
PC	Personal Computer
PDOP	Position Dilution of Precision
PDP	Programable Data Processor
PPS	Precise Positioning Service
PRN	Pseudorandom Noise
RAIM	Receiver Autonomous Integrity Monitoring
RF	Radio Frequency
RINEX	Receiver Independent Exchange (format)
SA	Selective Availability
SD	Single-Difference
SERIES	Satellite Emission Range Inferred Earth Surveying

SLR	Satellite Laser Ranging
SPOT	Satellite Probatoire d'Observation de la Terre
SPS	Standard Positioning Service
SV	Space Vehicle
TD	Triple-Difference
TDOP	Time Dilution of Precision
TDT	Terrestrial Dynamic Time
TEC	Total Electron Content
TLM	Telemetry
TM	Trade Mark
TOPEX	(Ocean) Topography Experiment
TRANSIT	Time Ranging and Sequential
UERE	User Equivalent Range Error
USGS	U.S. Geological Survey
UT	Universal Time
UTC	Universal Time Coordinated
UTM	Universal Transverse Mercator (projection)
VDOP	Vertical Dilution of Precision
VHSIC	Very High Speed Integrated Circuit
VLBI	Very Long Baseline Interferometry
WGS	World Geodetic System

1. Introduction

1.1 The origins of surveying

Since the dawn of civilization, man has looked to the heavens with awe searching for portentous signs. Some of these men became experts in deciphering the mystery of the stars and developed rules for governing life based upon their placement. The exact time to plant the crops was one of the events that was foretold by the early priest astronomers who in essence were the world's first surveyors. Today, we know that the alignment of such structures as the pyramids and Stonehenge was accomplished by celestial observations and that the structures themselves were used to measure the time of celestial events such as the vernal equinox. The chain of technical developments from these early astronomical surveyors to the present satellite geodesists reflects man's desire to be able to master time and space and to use science to further his society.

The surveyor's role in society has remained unchanged from the earliest days; that is to determine land boundaries, provide maps of his environment, and control the construction of public works.

Some of the first known surveyors were Egyptian surveyors who used distant control points to replace property corners destroyed by the flooding Nile River.

Surveys on a larger scale were conducted by the French surveyors Cassini and Picard, who measured the interior angles of a series of interconnecting triangles in combination with measured baselines, to determine the coordinates of points extending from Dunkirk to Collioure. The triangulation technique was subsequently used by surveyors as the main means of determining accurate coordinates over continental distances.

1.2 Development of global surveying techniques

The use of triangulation (later combined with trilateration and traversing) was limited by the line-of-sight. Surveyors climbed to mountain tops and developed special survey towers to extend this line-of-sight usually by small amounts. The series of triangles was generally oriented or fixed by astronomic points where special surveyors had observed selected stars to determine the position of that point on the surface of the earth. Since these astronomic positions could be in error by hundreds of meters, each continent was virtually (positionally) isolated and their interrelationship was imprecisely known.

1.2.1 Optical global triangulation

Some of the first attempts to determine the interrelationship between the continents were made using the occultation of certain stars by the moon. This method was cumbersome at best and was not particularly successful. The launch of the Russian Sputnik satellite in 1957, however, had tremendously advanced the connection of the various world datums. In the beginning of the era of artificial satellites, an optical method, based (in principle) on the stellar triangulation method developed in Finland as early as 1946, was applied very successfully. The worldwide satellite triangulation program often called the BC-4 program (after the camera that was used) for the first time determined the interrelationships of the major world datums. This method involved photographing special reflective satellites against a star background with a metric camera that was fitted with a specially manufactured chopping shutter. The image that appeared on the photograph consisted of a series of dots depicting each star's path and a series of dots depicting the satellite's path. The coordinates of selected dots were precisely measured using a photogrammetric comparator, and the associated spatial directions from the observing site to the satellite were then processed using an analytical photogrammetric model. Photographing the same satellite from a neighbouring site simultaneously and processing the data in an analogous way yields another set of spatial directions. Each pair of corresponding directions forms a plane containing the observing points and the satellite and the intersection of at least two planes results in the spatial direction between the observing sites. In the next step, these oriented directions were used to construct a global network such that the scale was derived from several terrestrial traverses. An example is the European baseline running from Tromsø in Norway to Catania on Sicily. The main problem in using this optical technique was that clear sky was required simultaneously at a minimum of two observing sites separated by some 4 000 km, and the equipment was massive and expensive. Thus, optical direction measurement was soon supplanted by the electromagnetic ranging technique because of all-weather capability, greater accuracy, and lower cost of the newer technique.

1.2.2 Electromagnetic global trilateration

First attempts to (positionally) connect the continents by electromagnetic techniques was by the use of HIRAN, an electronic HIgh RANging system developed during World War II to position aircraft. Beginning in the late

1940's, HIRAN arcs of trilateration were measured between North America and Europe in an attempt to determine the difference in their respective datums. A significant technological breakthrough occurred when scientists around the world experienced that the Doppler shift in the signal broadcast by a satellite could be used as an observable to determine the exact time of closest approach of the satellite. This knowledge, together with the ability to compute satellite ephemerides according to Kepler's laws, led to the present capability of instantaneously determining precise position anywhere in the world.

The immediate predecessor of today's modern positioning system is the Navy Navigational Satellite System (NNSS), also called TRANSIT system. This system was composed of six satellites orbiting at altitudes of about 1100 km with nearly circular polar orbits. The TRANSIT system was developed by the U.S. military, primarily, to determine the coordinates of vessels and aircraft. Civilian use of this satellite system was eventually authorized and the system became widely used worldwide both for navigation and surveying. Today, thousands of small vessels and aircraft use the TRANSIT system to determine their position worldwide.

Some of the early TRANSIT experiments by the U.S. Defense Mapping Agency (DMA) and the U.S. Coast & Geodetic Survey showed that accuracies of about one meter could be obtained by occupying a point for several days and reducing the observations using the postprocessed precise ephemerides. Groups of Doppler receivers in translocation mode could also be used to determine the relative coordinates of points to submeter accuracy using the broadcast ephemerides. This system employed essentially the same Doppler observable used to track the Sputnik satellite; however, the orbits of the TRANSIT satellites were precisely determined by tracking them at widely spaced fixed sites. The TRANSIT satellites are still being used to determine the coordinates of selected datum points.

1.3   History of the Global Positioning System

The Global Positioning System (GPS) was developed to replace the TRANSIT system because of two major shortcomings in the earlier system. The main problem with TRANSIT was the large time gaps in coverage. Since nominally a satellite would pass overhead every 90 minutes, users had to interpolate their position between "fixes" or passes. The second problem with the TRANSIT system was its relatively low navigation accuracy.

In contrast, GPS answers the questions "What time, what position, and what velocity is it?" quickly, accurately and inexpensively anywhere on the globe at any time, cf. Remondi (1991c).

1.3.1 Navigating with GPS

The aim of navigation is instantaneous positioning and velocity determination. As stated, one of the main problems with the TRANSIT system was the fact that the seven orbiting satellites were not able to provide continuous positioning.

Satellite constellation. To provide a continuous global positioning capability, a scheme to orbit a sufficient number of satellites to ensure that four were always electronically visible was developed for GPS. Several schemes were proposed and it was found that 21 evenly spaced satellites placed in circular 12-hour orbits inclined 55° to the equatorial plane would provide the desired coverage for the least expense. In any event, the planned constellation will provide a minimum of four satellites in good geometric position 24 hours per day anywhere on the earth. Depending on the selected elevation angle there will often be more than the minimum number of satellites available for use and it is during these periods that surveyors will perform kinematic and other special surveys. In fact, assuming a 10° elevation angle there are brief periods where up to 10 GPS satellites are visible on the earth.

Point positioning. The GPS satellites are configured, primarily, to provide the user with the capability of determining his position, expressed for example by latitude, longitude, and elevation. This is accomplished by the simple resection process using the distances measured to satellites.

Consider the satellites frozen in space at a given instant. The space coordinates $\underline{\varrho}^S$ relative to the center of the earth of each satellite can be computed from the ephemerides broadcast by the satellite by an algorithm presented in Chap. 4. If the ground receiver defined by the geocentric position vector $\underline{\varrho}_R$ employed a clock that was set precisely to GPS system time, cf. Sect. 3.3, the true distance or range ϱ to each satellite could be accurately measured by recording the time required for the (coded) satellite signal to reach the receiver. Each range defines a sphere with the center in the satellite for the location of the receiver. Hence, using this technique, ranges to only three satellites would be needed since the intersection of three spheres yields the three unknowns (e.g., latitude, longitude, and height) and could be determined from the three range equations, cf. also Fig. (1.1),

$$\varrho = \|\underline{\varrho}^S - \underline{\varrho}_R\|. \tag{1.1}$$

Modern GPS receivers apply a slightly different technique. They typically use an inexpensive crystal clock which is set approximately to GPS time. The clock of the ground receiver is thus offset from true GPS time, and because

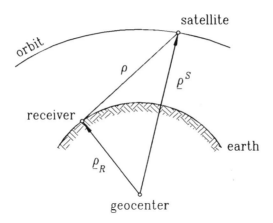

Fig. 1.1. Principle of satellite positioning

of this offset, the distance to the satellite is slightly longer or shorter than the "true" range. The receiver can overcome this problem by measuring four distances to four satellites (simultaneously). These distances are called pseudoranges R since they are the true range plus (or minus) a small extra distance $\Delta\varrho$ resulting from the receiver clock error or bias δ. A simple model for the pseudorange is

$$R = \varrho + \Delta\varrho = \varrho + c\,\delta \tag{1.2}$$

with c being the velocity of light.

The point position can be solved by resection as before except we now need four pseudoranges to solve for the four unknowns; these are three components of position plus the clock bias. It is worth noting that the range error $\Delta\varrho$ could be eliminated in advance by differencing the pseudoranges measured from one site to two satellites or two different positions of one satellite. In the second case, the resulting delta range corresponds to the observable in the TRANSIT system. In both cases, the delta range now defines a hyperboloid with its foci placed in the two satellites or the two different satellite positions for the geometrical location of the receiver.

Considering the fundamental observation equation (1.1), one can conclude that the accuracy of the position determined using a single receiver essentially is affected by the following factors:

- Accuracy of each satellite's position.

- Accuracy of pseudorange measurement.

- Geometry.

Systematic errors in the satellite's position and eventual satellite clock biases in the pseudoranges can be eliminated by differencing the pseudoranges measured from two sites to the satellite. This interferometric approach has become fundamental for GPS surveying as demonstrated below. However, no mode of differencing can overcome poor geometry.

A measure for satellite geometry with respect to the observing site is a factor known as Geometric Dilution of Precision (GDOP). In a geometric approach this factor is inversely proportional to the volume of a body formed by the top of unit vectors between the observing site and the satellites. More details and an analytical approach on this subject are provided in Sect. 9.5.

Velocity determination. The determination of the instantaneous velocity of a moving vehicle is another goal of navigation. This can be achieved by using the Doppler principle of radio signals. Because of the relative motion of the GPS satellites with respect to a moving vehicle, the frequency of a signal broadcast by the satellites is shifted when received at the vehicle. This measurable Doppler shift is proportional to the relative radial velocity. Since the radial velocity of the satellites is known, the radial velocity of the moving vehicle can be deduced from the Doppler observable.

In summary, GPS was designed to solve many of the problems inherent to the TRANSIT system. Above all, GPS (in its final stage) will provide 24 hours a day instantaneous global navigation to positioning accuracies of a few meters. However, the system as originally designed did not include provision for the accurate surveying that is performed today. This surveying use of GPS resulted from a number of fortuitous developments described below.

1.3.2   Surveying with GPS

From navigation to surveying. As previously described, the use of near earth satellites for navigation was demonstrated by the TRANSIT system. In 1964, I. Smith filed a patent describing a satellite system that would emit time codes and radio waves that would be received on earth as time delayed transmissions creating hyperbolic lines of position, cf. Smith (1964). This concept would become important in the treatment of GPS observables to compute precise vectors. A few years later, another patent was filed by R. Easton further refining the concept of comparing the phase from two or more satellites, cf. Easton (1970).

In 1972, C. Counselman along with his colleagues from the Massachusetts Institute of Technology's (MIT) Department of Earth and Planetary Sciences reported on the first use of interferometry to track the Apollo 16 Lunar Rover module, cf. Counselman et al. (1972). The principle they described is in essence the same technique they used later in developing the first geodetic GPS receiver and corresponds to differencing pseudoranges measured from two receivers to one satellite. The present use of the GPS carrier phase to make millimeter vector measurements dates from work by the MIT group using Very Long Baseline Interferometry (VLBI) performed between 1976 and 1978 where they proved millimeter accuracy was obtainable using the interferometric technique, cf. Rogers et al. (1978).

The present GPS survey system is essentially described in a paper by Counselman and Shapiro (1978). The Miniature Interferometer Terminals for Earth Surveying (MITES) detail how a satellite system can be used for precise surveying. This concept was further refined to include the NAV-STAR system in a NASA paper authored by Counselman et al. (1979). This paper also presents a description of the codeless technique that later became important in developing high-accuracy dual frequency receivers. The main significance of the MIT group's contribution to GPS is they demonstrated for the first time that the GPS carrier signal could be processed by differencing the phases, so that vectors between two points could be measured to millimeter (for short lines) accuracy.

Observation techniques. It should be noted that when we refer to high accuracy GPS surveying we refer to the precise measurement of the vector between two (or more) GPS instruments. The observation technique where both receivers involved remain fixed in position is called "static" surveying. The static method formerly required hours of observation and was the technique that was primarily used for early GPS surveys.

A second technique where one receiver remains fixed while the second receiver moves is called "kinematic" surveying. Remondi (1986) first demonstrated that subcentimeter vector accuracies could be obtained between a pair of GPS survey instruments with as little as a few seconds of data collection using this method. Surveys are performed by first placing a pair of receivers on two points of known location where data are collected from four (or preferably more) satellites for several minutes. Following this brief initialization, one of the receivers can be moved and as long as four or more satellites are continuously (no loss of lock) tracked, the vector between the fixed and roving instruments can be determined to high accuracies. Remondi also reported a variation of this technique to quickly determine the initial vector for kinematic surveys. This variant has been termed the antenna swap technique since the instruments are swapped between two points at the beginning of the survey to determine the initial vector, cf. Hofmann-Wellenhof and Remondi (1988). An antenna swap can be performed in approximately 1 minute.

Remondi (1988) also first developed another survey technique that is a variant of the normal static method. In this "pseudokinematic" technique a pair of receivers occupies a pair of points for two brief (e.g., 2–5 minutes) periods that are separated in time (e.g., 30–60 minutes). This method has also been called intermittent static or snapshot static and has demonstrated accuracies comparable to the static method.

The "differential" positioning technique involves placing a continuous tracking receiver at a fixed site of known position. Comparing computed pseudoranges with measured pseudoranges the reference site can transmit corrections to a roving receiver to improve its measured pseudoranges. This technique provides real-time accurate positioning at the one-meter level. Hence, it meets many requirements dictated by the complexity of modern civilization.

Hardware developments. The following sections contain reference to various terms that are more fully described in subsequent chapters. These are the C/A-code (Coarse/Acquisition) and P-code (Precision) which are basically code bits that are modulated on the two carrier signals broadcast by the GPS satellites. Code correlation as well as codeless techniques strip these codes from the carrier so that the phase of the (reconstructed) carrier can be measured. Brand names mentioned in this section are included for historical purposes since they represent the first of a certain class or type of receiver.

An interferometric technology for codeless pseudoranging was developed by P. MacDoran at the California Institute of Technology, Jet Propulsion Laboratory (JPL), with financial support from the National Aeronautics and Space Administration (NASA). This SERIES (Satellite Emission Range Inferred Earth Surveying) technique was later improved for commercial geodetic applications, cf. MacDoran et al. (1985). The culmination of the VLBI interferometric research applied to earth orbiting satellites was the production of a "portable" codeless GPS receiver that could measure short baselines to millimeter accuracy and long baselines to one part per million (ppm), cf. Collins (1982). This receiver trade-named the Macrometer Interferometric Surveyor^{TM} (Macrometer is a trademark of Aero Service Division, Western Atlas International, Houston, Texas) was tested by the U.S. Federal Geodetic Control Committee (FGCC), cf. Hothem and Fronczek (1983), and was used shortly thereafter in commercial surveys.

A parallel development was being carried out by the DMA in cooperation with the U.S. National Geodetic Survey (NGS) and the U.S. Geological Survey (USGS). In 1981, these agencies developed specifications for a portable dual frequency code correlating receiver that could be used for precise surveying and point positioning. Texas Instruments Company was awarded the contract to produce a receiver later trade-named the TI-4100. The NGS participated in developing specifications for the TI-4100 and their geodesists, C. Goad and B. Remondi, developed software to process its carrier phase data in a manner similar to the method used by the MIT group (i.e., interferometrically).

The physical characteristics of the TI-4100 were significantly different from the Macrometer. The TI-4100 was a dual frequency receiver that used the P-code to track a maximum of four satellites, while the original Macrometer was a rack mounted codeless single frequency receiver that simultaneously tracked up to six satellites. There were also significant logistical differences in performing surveys using these two pioneer instruments. The TI-4100 received the broadcast ephemerides and timing signals from the GPS satellites so units could be operated independently while the Macrometer required that all units be brought together prior to the survey and after the survey so that the time of the units could be synchronized. Also, the Macrometer required that the ephemerides for each day's tracking be generated at the home office prior to each day's observing session.

The next major development in GPS surveying occurred in 1985 when manufacturers started to produce C/A-code receivers that measured and output the carrier phase. The first of this class of receivers was trade-named the Trimble 4000S. This receiver required the data to be collected on an external (i.e., laptop) computer. The 4000S was the first of the generic C/A-code receivers that eventually were produced by a host of manufacturers. The first Trimble receivers were sold without processing software; however, the company soon retained the services of C. Goad who produced appropriate vector computation software which set the standard for future software developers.

Today's GPS receivers include all features of the early models and additionally have expanded capabilities. By far the major portion of receivers produced today are the C/A-code single frequency type. However, for precise geodetic work dual frequency receivers are becoming the standard. Many survey receivers now have incorporated the codeless technology to track the second frequency, and other receivers use all three techniques (C/A-code, P-code, codeless) to track satellites on both broadcast frequencies. These advanced receivers provide the greatest accuracy and productivity although they are more expensive than the simpler C/A-code receivers.

Software developments. The development of GPS surveying software has largely paralleled the development of hardware. Most of the receivers that can be used for surveying are sold with a suite of personal computer (PC) programs that use the carrier phase data to compute the vectors between occupied points.

The NGS has been one of the primary organizations in the world in developing independent GPS processing software. As previously mentioned, C. Goad and B. Remondi pioneered the development of receiver independent software.

The NGS first produced processing software that used the Macrometer phase measurements and the precise ephemerides produced by the U.S. Naval Surface Warfare Center (NSWC). Other Macrometer users had to apply the processing software developed by the Macrometer manufacturer which required the use of specially formatted ephemerides produced (and sold) by them. The NGS software was also adapted for the TI-4100 format data and finally for other receivers that were subsequently used.

The original software developed by both the NGS and manufacturers computed individual vectors one at a time. These vectors were then combined in a network or geometric figure and the coordinates of all points were determined using least squares adjustment programs.

The NGS and the Macrometer manufacturer eventually developed processing software that simultaneously determined all vectors observed during a given period of time (often called session). First, this software fixed the satellite positions in the same way as the vector by vector software. The second generation multibaseline software included the ability to determine corrections to the satellite orbits and is often called orbital relaxation software. This technique was pioneered by G. Beutler's group at the Bernese Astronomical Institute. Although today the majority of surveyors use the single vector computation software run in a "batch" computer mode, the orbit relaxation software is used for special projects requiring the highest accuracy (e.g., 0.01 ppm). Some GPS experts feel that the orbit relaxation software will be used in the future by land surveyors as well as geodesists to provide high accuracy surveys referenced to distant fixed tracking sites.

Ephemerides service. The first GPS surveys performed in late 1982 using Macrometers depended on orbital data derived from a private tracking network. Later, the broadcast ephemerides were used to supplement this private tracking data. The TI-4100 receiver obtained the ephemerides broadcast by the satellites so that processing programs could use this ephemerides to process vectors. The NSWC originally processed the military ephemerides, cf. Swift (1985), obtaining "precise" postprocessed ephemerides which was turned over to NGS for limited distribution to the public.

Today, the NGS in cooperation with various organizations around the world provide satellite tracking data from points that are referenced to the global VLBI network. These CIGNET (Cooperative International GPS Network) tracking stations collect code range and phase data for both frequencies for all satellites. These data are sent to NGS on a daily basis and are available to the public upon request. Theoretically, one could compute highly accurate orbits from this data set (using appropriate software).

The NGS now computes and distributes precise orbital data to the public. It is anticipated that the orbits will be at least two weeks old to satisfy U.S. Department of Defense (DoD) requirements; however, it may be theoretically possible but problematic to compute predicted orbits (from the two-week old data) that are nearly as accurate as the present broadcast ephemerides.

2. Overview of GPS

2.1 Basic concept

The Global Positioning System is the responsibility of the Joint Program Office (JPO) located at the U.S. Air Force Systems Command's Space Division, Los Angeles Air Force Base (AFB). In 1973, the JPO was directed by the U.S. Department of Defense (DoD) to establish, develop, test, acquire, and deploy a spaceborne positioning system. The present NAVigation System with Timing And Ranging (NAVSTAR) Global Positioning System (GPS) is the result of this initial directive.

The Global Positioning System was conceived as a ranging system from known positions of satellites in space to unknown positions on land, sea, in air and space. Effectively, the satellite signal is continually marked with its (own) transmission time so that when received the signal transit period can be measured with a synchronized receiver. Apart from point positioning, the determination of a vehicle's instantaneous position and velocity (i.e., navigation), and the precise coordination of time (i.e., time transfer) were original objectives of GPS. A definition given by Wooden (1985) reads:

"The Navstar Global Positioning System (GPS) is an all-weather, space-based navigation system under development by the Department of Defense (DoD) to satisfy the requirements for the military forces to accurately determine their position, velocity, and time in a common reference system, anywhere on or near the Earth on a continuous basis."

Since the DoD is the initiator of GPS, the primary goals were military ones. But the U.S. Congress, with guidance from the President, directed DoD to promote its civil use. This was greatly accelerated by employing the Macrometer for geodetic surveying. This instrument was in commercial use at the time the military was still testing navigation receivers so that the first productive application of GPS was to establish high-accuracy geodetic networks.

As previously stated, cf. Sect. 1.3.1, GPS uses pseudoranges derived from the broadcast satellite signal. The pseudorange is derived either from measuring the travel time of the (coded) signal and multiplying it by its velocity or by measuring the phase of the signal. In both cases, the clocks of the receiver and the satellite are employed. Since these clocks are never perfectly synchronized, instead of true ranges "pseudoranges" are obtained where the synchronization error (denoted as clock error) is taken into account, cf. Eq. (1.2). Consequently, each equation of this type comprises four unknowns: the desired three point coordinates contained in the true range, and the clock error. Thus, four satellites are necessary to solve for the four unknowns. Indeed, the GPS concept assumes that, when fully deployed, four or more satellites will be in view at any location on earth 24 hours a day. The solution becomes more complicated when using the measured phase. This observable is ambiguous by an integer number of signal wavelengths, so that the model for phase pseudoranges is augmented by an initial bias, also called integer ambiguity.

The all-weather global system managed by the JPO consists of three segments: (1) The space segment consisting of satellites which broadcast signals, (2) the control segment steering the whole system, and (3) the user segment including the many types of receivers.

2.2 Space segment

2.2.1 Constellation

When fully deployed, the space segment will provide global coverage with four to eight simultaneous observable satellites above 15° elevation. This is accomplished by satellites in nearly circular orbits with an altitude of about 20 200 km above the earth and a period of approximately 12 sidereal hours, cf. Perreault (1980); Rutscheidt and Roth (1982). This constellation and the number of satellites used have evolved from earlier plans for a 24-satellite and 3-orbital plane constellation, inclined 63° to the equator, cf. Mueller and Archinal (1981). Later, for budgetary reasons, the space segment was reduced to 18 satellites, with three satellites in each of six orbital planes. This scheme was eventually rejected, since it did not provide the desired 24-hour worldwide coverage. In about 1986, the number of satellites planned was increased to 21, again three each in six orbital planes, and three additional active spares, cf. Wells et al. (1987). The most recent plan calls for 21 operational satellites plus three active spares deployed in six planes with an inclination of 55° and with four satellites per plane. In this plan, the spare satellites are used to replace a malfunctioning "active" satellite. Consequently, three replacements are possible before one of the seven ground spares must be launched to maintain the full constellation, cf. Brunner (1984).

2.2.2 Satellites

General remarks. The GPS satellites, essentially, provide a platform for radio transceivers, atomic clocks, computers, and various ancillary equipment used to operate the system. The electronic equipment of each satellite allows the user to measure a pseudorange R to the satellite, and each satellite broadcasts a message which allows the user to determine the spatial position ϱ^S of the satellite for arbitrary instants. Given these capabilities, users are able to determine their position $\underline{\varrho}_R$ on or above the earth by resection, cf. Fig. 1.1. The auxiliary equipment of each satellite, among others, consists of two $7\,m^2$ solar panels for power supply and a propulsion system that enables orbit adjustments and stability control, cf. Payne (1982).

Satellite categories. There are three classes or types of GPS satellites. These are the Block I, Block II, and Block IIR satellites, cf. Jones (1989).

Eleven Block I satellites (weighing 845 kg) were launched by JPO in the period between 1978 to 1985 from Vandenberg AFB, California, with Atlas F launch vehicles. With the exception of one booster failure in 1981, cf. Jones (1989), all launches were successful. In March 1992, still five of the original Block I satellites remained in operation including one that was launched in 1978. This is remarkable since the 4.5 year design life of Block I satellites, cf. Stein (1986), has been surpassed for some of the satellites by a factor nearly three. The Block I constellation is slightly different from the Block II constellation since the inclination of their orbital planes is 63° compared to the 55° inclination in the more recent plans.

The 28 Block II satellites presently being manufactured are designated for the first operational constellation, cf. Jones (1989). Of this total number, 21 active and three spares will be deployed. The first Block II satellite, costing approximately $ 50 million and weighing more than 1 500 kg, was launched on February 14, 1989 from the Kennedy Space Center, Cape Canaveral AFB in Florida, using Delta II Rockets, cf. Stein (1986). The mean mission duration of the Block II satellites is six years, and their design goal is 7.5 years. Individual satellites can easily remain operational as long as 10 years since their consumables will last this long, cf. Payne (1982). An important difference between Block I and Block II satellites relates to U.S. national security. Block I satellite signals were fully available to civilian users while some Block II signals are restricted.

The GPS satellites which will replace the Block II's are the Block IIR's which have a design life of 10 years. The "R" denotes replenishment or replacement. These satellites are currently under development, with the first satellites planned for delivery by 1995. The Block IIR's are expected to have on-board hydrogen masers. These atomic clocks are at least one order of magnitude more precise than the atomic clocks in the Block II satellites. The Block IIR satellites will also have improved facilities for communication and improved on-board orbit capability, since intersatellite tracking is provided. The Block IIR's weigh more than 2 000 kg but are only one-half of the cost of the Block II's, cf. Montgomery (1991). It was planned to orbit the Block IIR satellites using the Space Shuttle. Each Shuttle would be capable of transporting up to three satellites so that rapid deployment of the constellation is possible. However, these plans may change as the space program develops.

Satellite signal. The actual carrier broadcast by the satellite is a spread spectrum signal that makes it less subject to intentional (or unintentional) jamming. The spread spectrum technique is commonly used today by such diverse equipment as hydrographic positioning ranging systems and wireless Local Area Network (LAN) systems.

The key to the system's accuracy is the fact that all signal components are precisely controlled by atomic clocks. The Block II satellites have four on-board time standards, two rubidium and two cesium clocks. The long-term frequency stability of these clocks reaches a few parts in 10^{-13} and 10^{-14} over one day. The hydrogen masers planned for the Block IIR's have a stability of 10^{-14} to 10^{-15} over one day, cf. Scherrer (1985). These highly accurate frequency standards being the heart of GPS satellites produce the fundamental L-band frequency of 10.23 MHz. Coherently derived from this fundamental frequency are two signals, the $L1$ and the $L2$ carrier waves generated by multiplying the fundamental frequency by 154 and 120, respectively, thus yielding

$$L1 = 1575.42 \text{ MHz}$$
$$L2 = 1227.60 \text{ MHz}.$$

These dual frequencies are essential for eliminating the major source of error, i.e., the ionospheric refraction, cf. Sect. 6.3.2.

The pseudoranges that are derived from measured travel time of the signal from each satellite to the receiver use two pseudorandom noise (PRN) codes that are modulated (superimposed) onto the two base carriers.

The first code is the C/A-code (Coarse/Acquisition-code), also designated as the Standard Positioning Service (SPS), which is available for civilian use. The C/A-code with an effective wavelength of approximately 300 m is modulated only upon $L1$ and is purposely omitted from $L2$. This omission allows the JPO to control the information broadcast by the satellite, and thus denies full system accuracy to nonmilitary users.

The second code is the P-code (Precision-code), also designated as the Precise Positioning Service (PPS), which has been reserved for use by the U.S. military and other authorized users. The P-code with an effective wavelength of approximately 30 m is modulated on both carriers $L1$ and $L2$.

Present JPO policy is to permit unlimited access to the P-code until such time as the system is declared fully operational.

In addition to the PRN codes a data message is modulated onto the carriers comprising: satellite ephemerides, ionospheric modeling coefficients, status information, system time and satellite clock bias, and drift information. A detailed signal description is given in Sect. 5.1.

Satellite identification. The satellites have various systems of identification: launch sequence number, orbital position number, assigned vehicle PRN code, NASA catalogue number, and international designation. To avoid any confusion, only the PRN number is used, which is also taken for the satellite navigation message, cf. Wells (1985).

Satellite configuration. With the full constellation, four to eight satellites (above 15° elevation) can be observed simultaneously from anywhere on earth at any time of day. If the elevation mask is reduced to 10°, occasionally up to 10 satellites will be visible; and if the elevation mask is further reduced to 5°, occasionally 12 satellites will be visible. Until the full constellation is deployed, the usefulness of GPS will be restricted to a portion of the day depending upon the user's location. For the present, some surveying may still have to be performed during the hours of darkness. In the U.S. there is presently more time available for measurement than can be conveniently used by one survey crew. These "windows" of satellite availability are shifted by four minutes each day, due to the difference between sidereal time and Universal Time (UT). For example, if some of the satellites appeared in a given geometric configuration at 9:00 UT today, they would be roughly in the same position in the sky at 8:56 UT the following day.

2.2.3 Denial of accuracy and access

There are basically two methods for denying civilian users full use of the system. The first is Selective Availability (SA) and the second method is Anti-spoofing (A-S).

Selective availability. Primarily, this kind of denial has been accomplished by "dithering" the satellite clock frequency in a way that prevents civilian users from accurately measuring instantaneous pseudoranges. This form of accuracy denial mainly affects any one-receiver operation. When pseudoranges are differenced between two receivers, the dithering effect is largely eliminated, so that this navigation mode proposed for example by the U.S. Coast Guard will remain unaffected. The SA has only been implemented in

Block II satellites and has been in force intermittently since April 1990 at various levels of accuracy denial.

The second method of accuracy denial is to truncate the transmitted navigation message so that the coordinates of the satellites cannot be accurately computed. The error in satellite position roughly translates to a like position error of the receiver.

Anti-spoofing. The design of the GPS system includes the ability to essentially "turn off" the P-code or invoke an encrypted code (Y-code) as a means of denying access to the P-code to all but authorized users. The rationale for doing this is to keep adversaries from sending out false signals with the GPS signature to create confusion and cause users to misposition themselves. Under present policy, the A-S is scheduled to be activated when the system is fully operational. When this is done, access to the P-code is only possible by installing in each receiver channel an Auxiliary Output Chip (AOC) which will be available only on an authorized basis. Thus, A-S will affect many of the high accuracy survey uses of the system.

2.3   Control segment

This segment comprises the Operational Control System (OCS) which consists of a master control station, worldwide monitor stations, and ground control stations. The main operational tasks of the control segment are: tracking of the satellites for the orbit and clock determination and prediction modeling, time synchronization of the satellites, and upload of the data message to the satellites. There are many nonoperational activities, such as procurement and launch, that will not be addressed here.

2.3.1   Master control station

The location of the master control station was formerly at Vandenberg AFB, California, but has been moved to the Consolidated Space Operations Center (CSOC) at Falcon AFB, Colorado Springs, Colorado. CSOC collects the tracking data from the monitor stations and calculates the satellite orbit and clock parameters using a Kalman estimator. These results are then passed to one of the three ground control stations for eventual upload to the satellites. The satellite control and system operation is also the responsibility of the master control station.

2.3.2 Monitor stations

There are five monitor stations located at: Hawaii, Colorado Springs, Ascension Island in the South Atlantic Ocean, Diego Garcia in the Indian Ocean, and Kwajalein in the North Pacific Ocean, cf. Gouldman et al. (1989). Each of these stations is equipped with a precise cesium time standard and P-code receivers which continuously measure the P-code pseudoranges to all satellites in view. Pseudoranges are tracked every 1.5 seconds and using the ionospheric and meteorological data, they are smoothed to produce 15-minute interval data which are transmitted to the master control station.

The tracking network described above is the official network for determining the broadcast ephemerides as well as modeling the satellite clocks. For the precise ephemerides the data of five additional (DMA) sites are used. Other private tracking networks do exist, however. These private networks generally determine the ephemerides of the satellites after the fact and have no part in managing the system. One such private tracking network has been operated by the manufacturer of the Macrometer since 1983. Another more globally oriented tracking network is the Cooperative International GPS Network (CIGNET). This network is being operated by the NGS with tracking stations located at VLBI sites. More details on this network are provided in Sect. 4.4.1.

2.3.3 Ground control stations

These stations collocated with the monitor stations at Ascension, Diego Garcia, and Kwajalein, cf. Bowen et al. (1986), are the communication links to the satellites and mainly consist of the ground antennas. The satellite ephemerides and clock information, calculated at the master control station and received via communication links, are uploaded to each GPS satellite via S-band radio links, cf. Rutscheidt and Roth (1982). Formerly, uploading to each satellite was performed every eight hours, cf. Stein (1986), at present the rate has been reduced to once per day, cf. Remondi (1991b). If a ground station becomes disabled, prestored navigation messages are available in each satellite to support a 14-day prediction span that gradually degrades positioning accuracy from 10 to 200 meters.

2.4 User segment

2.4.1 User categories

Military user. Strictly speaking, the term "user segment" is related to the U.S. DoD concept of GPS as an adjunct to the national defense program.

Even during the early days of the system, it was planned to incorporate a GPS receiver into virtually every major defense system. It was envisioned that every aircraft, ship, land vehicle, and even groups of infantry would have an appropriate GPS receiver to coordinate their military activities. In fact, many GPS receivers were used as planned during the 1991 Gulf War under combat conditions. In this war, the SA which had been previously invoked was turned off so that troops could use more readily available civilian receivers. Hand held C/A-code receivers were particularly useful in navigating the featureless desert.

There are various other military uses that have recently been proposed. One manufacturer offers a receiver that can be connected to four antennas. When the antennas are placed in a fixed array (e.g., corners of a square), the three rotation angles of the array can be determined in addition to its position. For example, placing antennas on the bow, stern, and port and starboard points of a ship would result in the determination of pitch, roll, yaw, and position of the vessel.

More unconventional uses of the system include measuring the position of lower altitude satellites. On-board GPS receivers will permit satellites to be accurately positioned with much less effort than the present tracking techniques. For example, the utility of imaging satellites such as the French SPOT satellite would be greatly improved by an on-board GPS receiver to determine the precise position of each image.

Civilian user. The civilian use of GPS occurred several years ahead of schedule in a manner not envisioned by the system's planners. The primary focus in the first few years of the system's development was on navigation receivers. As previously described in Chap. 1, the SERIES technique at JPL and the development of the Macrometer by C. Counselman started the GPS surveying revolution. The primary concept of using an interferometric rather than Doppler solution model meant that GPS could be used for not only long line geodetic measurements but also for the most exacting short line land survey measurements.

Today, GPS receivers are being used to conduct all types of land and geodetic control surveys, and tests are being conducted to use the system to precisely position photo-aircraft to reduce the amount of ground control needed for mapping.

The nonsurveyor civilian uses of GPS will significantly outnumber the survey uses of the system. One of the major uses of GPS will be for fleet management and control. Several cities are equipping emergency vehicles with receivers and computers with screens that display the cities' road system. The location of each emergency vehicle can be sent to a dispatcher by radio link so that he can keep track of his resources and reroute vehicles when necessary. Similar systems are planned to track trains and freight hauling vehicles. Naturally, all aircraft and vessels will be equipped with GPS in the near future.

GPS is also being used by hikers and boaters to determine their locations. One Japanese manufacturer is presently offering a GPS/computer graphics system for use in automobiles at the cost of a good high fidelity music system.

2.4.2 Receiver types

The uses of GPS described in the previous section are just a sample of the applications of this system. The diversity of the uses is matched by the type of receivers available today. This section will give an overview of the equipment marketed today; however, more details will be provided in Sect. 5.2.2. Based on the type of observables (i.e., code pseudoranges or carrier phases) and on the availability of codes (i.e., C/A-code or P-code), one can classify GPS receivers into three groups: (1) C/A-code pseudorange, (2) C/A-code carrier phase, and (3) P-code carrier phase measuring instruments.

C/A-code pseudorange receivers. This type of receiver is usually a hand held device powered by flashlight batteries. Typical devices have from one to six independent receiver channels and output the three-dimensional position either in longitude, latitude, and height or in some map projection system (e.g., UTM coordinates and height). Receivers with four or more channels are preferred for applications where the receiver is in motion since simultaneous satellite ranges can be measured to produce more accurate positions. On the other hand, a single channel receiver is adequate for applications where the receiver is at a fixed location and the range measurements can be sequentially determined. The basic multichannel C/A-code pseudorange receiver is the type of receiver that will be used by hikers, boaters, and eventually in automobiles.

C/A-code carrier receivers. Most of the receivers for surveying during the period from 1985 to 1992 use the C/A-code to acquire and lock on to the $L1$ carrier. Most instruments have a minimum of four independent receiver channels and some of the more recent designs have 12 channels. These receivers perform all the functions of the previously described models, and in addition store the time-tagged code range and carrier phase in some type of memory. Early models used laptop computers and magnetic tapes to store the measured data. Later models store measurement data in memory chips.

This type of receiver has recently been augmented to measure the phases of the $L2$ carrier by the use of the (codeless) squaring technique. This is accomplished by multiplying the $L2$ signal by itself to recover the phase of the carrier at one-half its wavelength. The P-code modulated on the $L2$ carrier is lost in the process and the signal to noise ratio is considerably lower than the C/A-code $L1$ measurement. Normally, the $L2$ phase is used in combination with the $L1$ measurement to reduce the ionospheric effect on the signal and thus provide a more accurate vector determination (especially for long lines).

These receivers can be used for all types of precise surveys including static, kinematic, and pseudokinematic.

P-code receivers. This type of receivers uses the P-code and thus enables lock on $L1$ and $L2$ carrier. One of the first receivers for surveying, point positioning, and navigation was the P-code receiver TI-4100 completed in 1984. This receiver was developed more from a military perspective than a civilian one and only military-related development could have attempted this. Manufacturers of civilian receivers first were able to justify P-code work around 1989–1990. In the fall of 1991, the U.S. Federal Geodetic Control Committee (FGCC) tested another P-code receiver which was the first receiver to be tested since the earlier TI-4100 FGCC tests. This test demonstrated the two main advantages of the P-code receiver. The first is its capability to measure long (100 km) lines to an accuracy of a few centimeters. The second advantage is that P-code instruments can measure moderate length lines (20 km) to an accuracy of a few centimeters with as little as ten minutes of data using a technique known as wide-laning which is based on a linear combination of the measured phases of $L1$ and $L2$.

3. Reference systems

3.1 Introduction

Consider the basic observation equation which relates the range ϱ with the instantaneous position vector $\underline{\varrho}^S$ of a satellite and the position vector $\underline{\varrho}_R$ of the observing site:

$$\varrho = \|\underline{\varrho}^S - \underline{\varrho}_R\| . \tag{3.1}$$

In Eq. (3.1) both vectors must be expressed in a uniform coordinate system. The definition of a three-dimensional Cartesian system requires a convention for the orientation of the axes and for the position of the origin.

For global applications such as satellite geodesy, equatorial coordinate systems are appropriate. According to Fig. 3.1, a space-fixed or inertial system $\underline{X}_i^0$ and an earth-fixed or terrestrial system $\underline{X}_i$ must be distinguished where $i = 1, 2, 3$. The earth's rotational vector $\underline{\omega}_E$ serves as $\underline{X}_3$-axis in both cases. The $\underline{X}_1^0$-axis for the space-fixed system points towards the vernal equinox and is thus the intersection line between the equatorial and the ecliptic plane. The $\underline{X}_1$-axis of the earth-fixed system is defined by the intersection line of the equatorial plane with the plane represented by the Greenwich meridian. The angle Θ_0 between the two systems is called Green-

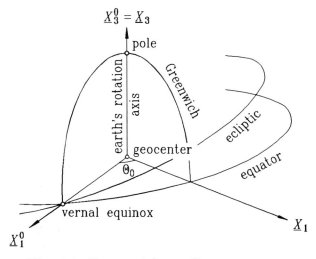

Fig. 3.1. Equatorial coordinate systems

wich sidereal time. The X_2-axis being orthogonal to both the X_1-axis and the X_3-axis completes a right-handed coordinate frame.

A coordinate system whose origin is located at the barycenter is at rest with respect to the solar system and thus conforms to Newtonian mechanics. In a geocentric system, however, accelerations are present because the earth is orbiting the sun. Thus, in such a system the laws of general relativity must be taken into account. But, since the main relativistic effect is caused by the gravity field of the earth itself, the geocentric system is better suited for the description of the motion of a close earth satellite. Note that the axes of a geocentric coordinate system remain parallel because the motion of the earth around the sun is described by revolution without rotation.

The earth's rotational vector denoted by $\underline{\omega}_E$ oscillates due to several reasons. The basic differential equations describing the oscillations follow from classical mechanics and are given by

$$\underline{M} = \frac{d\underline{N}}{dt} \tag{3.2}$$

$$\underline{M} = \frac{\partial \underline{N}}{\partial t} + \underline{\omega}_E \times \underline{N} \tag{3.3}$$

where $\underline{M}$ denotes a torque vector and $\underline{N}$ the angular momentum vector of the earth, cf. Moritz and Mueller (1988), Eqs. (2-54) and (2-59). The symbol "$\times$" indicates a vector or cross product. The torque $\underline{M}$ originates mainly from the gravitational forces of sun and moon, hence it is closely related to the tidal potential. Equation (3.2) is expressed in a (quasi-) inertial system such as $\underline{X}_i^0$ and Eq. (3.3) holds for the rotating system $\underline{X}_i$. The partial derivative expresses the temporal change of $\underline{N}$ with respect to the earth-fixed system and the vector product considers the rotation of this system with respect to the inertial system. The earth's rotational vector $\underline{\omega}_E$ is related to the angular momentum vector $\underline{N}$ by the inertia tensor $\underline{C}$ as

$$\underline{N} = \underline{C}\,\underline{\omega}_E \,. \tag{3.4}$$

Introducing for the earth's rotational vector $\underline{\omega}_E$ its unit vector $\underline{\omega}$ and its norm $\omega_E = \|\underline{\omega}_E\|$, the relation

$$\underline{\omega}_E = \omega_E \, \underline{\omega} \tag{3.5}$$

can be formed.

The differential equations (3.2) and (3.3) can be separated into two parts. The oscillations of $\underline{\omega}$ are responsible for the variations of the X_3-axis and are considered in the subsequent section. The oscillations of the norm ω_E cause variations in the speed of rotation which are treated in the section on time systems.

Considering only the homogeneous part ($\underline{M} = \underline{0}$) of Eqs. (3.2) and (3.3) leads to free oscillations. The inhomogeneous solution gives the forced oscillation. In both cases, the oscillations can be related to the inertial or to the terrestrial system. A further criterion for the solution concerns the inertia tensor. For a rigid earth and neglecting internal mass shifts, this tensor is constant; this is not the case for a deformable earth.

3.2 Coordinate systems

3.2.1 Definitions

Oscillations of axes. The oscillation of $\underline{\omega}$ with respect to the inertial space is called nutation. For the sake of convenience the effect is partitioned into the secular precession and the periodic nutation. The oscillation with respect to the terrestrial system is named polar motion. A simplified representation of polar motion is given by Fig. 3.2. The image of a mean position of $\underline{\omega}$ is denoted P in this polar plot of the unit sphere. The free oscillation results in a motion of the rotational axis along a circular cone, with its mean position as axis, and an aperture angle of about $0.''4 \approx 12\,\text{m}$. On the unit sphere, this motion is represented by a 6 m radius circle around P. The image of an instantaneous position of the free oscillating earth's rotational axis is denoted R_0. The period of the free motion amounts to about 430 days and is known as the Chandler period. The forced motion can also be described by a cone. In Fig. 3.2 this cone is mapped by the circle around the free position R_0. The radius of this circle is related to the tidal deformation and is approximately 0.5 m. The nearly diurnal period of the forced motion corresponds to the tesseral part of the tidal potential of second degree since the zonal and sectorial components have no influence.

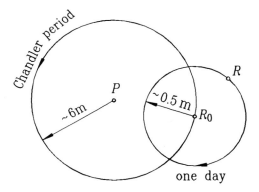

Fig. 3.2. Polar motion of earth's rotational axis

The respective motions of the angular momentum axis, which is within $0.''001$ of the rotation axis, are very similar. The free motion of the angular momentum axis deserves special attention because the forced motion can be removed by modeling the tidal attractions. The free polar motion is long-periodic and the free position in space is fixed since for $\underline{M} = \underline{0}$ the integration of Eq. (3.2) yields $\underline{N} = const$. By the way, this result implies the law of conservation of angular momentum as long as no external forces are applied. Because of the above mentioned properties, the angular momentum axis is appropriate to serve as a reference axis and the scientific community has named its free position in space Celestial Ephemeris Pole (CEP). A candidate for serving as reference axis in the terrestrial system is the mean position of the rotational axis denoted by P, cf. Fig. 3.2. This position is called Conventional International Origin (CIO). For historical reasons, the CIO represents the mean position of $\underline{\omega}$ during the period 1900 until 1905.

Conventional Inertial System. By convention, the $\underline{X}_3^0$-axis is identical to the position of the angular momentum axis at a standard epoch denoted by J2000.0, cf. Sect. 3.3. The $\underline{X}_1^0$-axis points to the associated vernal equinox. At present this equinox is realized kinematically by a set of fundamental stars, cf. Fricke et al. (1988). Since this system is defined conventionally and the practical realization does not necessarily coincide with the theoretical system, it is called Conventional Inertial Frame. Sometimes the term "quasi-inertial" is used to point out that a geocentric system is not rigorously inertial because of the accelerated motion of the earth around the sun.

Conventional Terrestrial System. Again by convention, the $\underline{X}_3$-axis is identical to the mean position of the earth's rotational axis as defined by the CIO. The $\underline{X}_1$-axis is associated with the mean Greenwich meridian. The realization of this system is named Conventional Terrestrial Frame and is defined by a set of terrestrial control stations serving as reference points, cf. for example Boucher and Altamimi (1989). Most of the reference stations are equipped with Satellite Laser Ranging (SLR) or Very Long Baseline Interferometry (VLBI) facilities.

Since 1987, GPS has used the World Geodetic System WGS-84 as a reference, cf. Decker (1986). Associated with WGS-84 is a geocentric equipotential ellipsoid of revolution which is defined by the four parameters listed in Table 3.1. However, using the theory of the equipotential ellipsoid, numerical values for other parameters such as the geometric flattening ($f = 1/298.2572221$) or the semiminor axis ($b = 6\,356\,752.314$ m) can be derived. Note that the parameter values have been adopted from the Geodetic Reference System 1980 (GRS-80) ellipsoid.

Table 3.1. Parameters of the WGS-84 ellipsoid

Parameter and Value	Explanation
$a \quad = 6\,378\,137$ m	Semimajor axis of ellipsoid
$J_2 \quad = 1\,082\,630 \cdot 10^{-9}$	Zonal coefficient of second degree
$\omega_E = 7\,292\,115 \cdot 10^{-11}$ rad $\cdot$ s^{-1}	Angular velocity of the earth
$\mu \quad = 3\,986\,005 \cdot 10^8$ m$^3 \cdot$ s^{-2}	Earth's gravitational constant

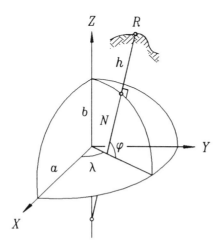

Fig. 3.3. Cartesian and ellipsoidal coordinates

A vector $\underline{X}$ in the terrestrial system can be represented by Cartesian coordinates X, Y, Z as well as by ellipsoidal coordinates φ, λ, h, cf. Fig. 3.3. The rectangular coordinates are often called Earth-Centered-Earth-Fixed (ECEF) coordinates. The relation between the two sets of coordinates is given by, cf. for example Heiskanen and Moritz (1967), p. 182,

$$
\underline{X} = \begin{bmatrix} X \\ Y \\ Z \end{bmatrix} = \begin{bmatrix} (N + h) \cos \varphi \, \cos \lambda \\ (N + h) \cos \varphi \, \sin \lambda \\ \left(\dfrac{b^2}{a^2} N + h \right) \sin \varphi \end{bmatrix} \tag{3.6}
$$

where φ, λ, h are the ellipsoidal latitude, longitude, height, N is the radius of curvature in prime vertical, and a, b are the semimajor and semiminor axis of ellipsoid. More details on the transformation of Cartesian and ellipsoidal (i.e., geodetic) coordinates are provided in Sect. 10.2.1.

3.2.2   Transformations

General remarks. The transformation between the Conventional Inertial System (CIS) and the Conventional Terrestrial System (CTS) is performed by means of rotations. For an arbitrary vector $\underline{x}$ the transformation is given by

$$\underline{x}_{[CTS]} = \underline{R}^M \, \underline{R}^S \, \underline{R}^N \, \underline{R}^P \, \underline{x}_{[CIS]} \tag{3.7}$$

with

$$
\begin{array}{lll}
\underline{R}^M & \ldots & \text{rotation matrix for polar motion} \\
\underline{R}^S & \ldots & \text{rotation matrix for sidereal time} \\
\underline{R}^N & \ldots & \text{rotation matrix for nutation} \\
\underline{R}^P & \ldots & \text{rotation matrix for precession.}
\end{array}
$$

The CIS, defined at the standard epoch J2000.0, is transformed into the instantaneous or true system at observation epoch by applying the corrections due to precession and nutation. The $\underline{X}_3^0$-axis of the true CIS represents the free position of the angular momentum axis and thus points to the CEP. Rotating this system around the $\underline{X}_3^0$-axis and through the sidereal time by the matrix $\underline{R}^S$ does not change the position of the CEP. Finally, the CEP is rotated into the CIO by $\underline{R}^M$ which completes the transformation.

The rotation matrices in Eq. (3.7) are composed of the elementary matrices $\underline{R}_i\{\alpha\}$ describing a positive rotation of the coordinate system around the $\underline{X}_i$-axis and through the angle α. As it may be verified from any textbook on vector analysis, the rotation matrices are given by

$$
\begin{aligned}
\underline{R}_1\{\alpha\} &= \begin{bmatrix} 1 & 0 & 0 \\ 0 & \cos\alpha & \sin\alpha \\ 0 & -\sin\alpha & \cos\alpha \end{bmatrix} \\[1em]
\underline{R}_2\{\alpha\} &= \begin{bmatrix} \cos\alpha & 0 & -\sin\alpha \\ 0 & 1 & 0 \\ \sin\alpha & 0 & \cos\alpha \end{bmatrix} \\[1em]
\underline{R}_3\{\alpha\} &= \begin{bmatrix} \cos\alpha & \sin\alpha & 0 \\ -\sin\alpha & \cos\alpha & 0 \\ 0 & 0 & 1 \end{bmatrix}.
\end{aligned}
\tag{3.8}
$$

Note that the matrices given by Eq. (3.8) are consistent with right-handed coordinate systems. The rotation angle α has a positive sign for clockwise rotation as viewed from the origin to the positive $\underline{X}_i$-axis.

Precession. A graphic representation of precession is given in Fig. 3.4. The position of the mean vernal equinox at the standard epoch t_0 is denoted by

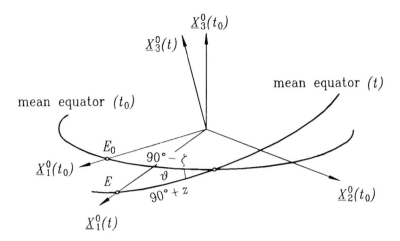

Fig. 3.4. Precession

E_0 and the position at the observation epoch t is denoted by E. The precession matrix $\underline{R}^P$ is composed of three successive rotation matrices

$$\underline{R}^P = \underline{R}_3\{-z\}\,\underline{R}_2\{\vartheta\}\,\underline{R}_3\{-\zeta\}$$

$$= \begin{bmatrix} \begin{array}{c} \cos z \cos \vartheta \cos \zeta \\ -\sin z \sin \zeta \end{array} & \begin{array}{c} -\cos z \cos \vartheta \sin \zeta \\ -\sin z \cos \zeta \end{array} & -\cos z \sin \vartheta \\[2ex] \begin{array}{c} \sin z \cos \vartheta \cos \zeta \\ +\cos z \sin \zeta \end{array} & \begin{array}{c} -\sin z \cos \vartheta \sin \zeta \\ +\cos z \cos \zeta \end{array} & -\sin z \sin \vartheta \\[2ex] \sin \vartheta \cos \zeta & -\sin \vartheta \sin \zeta & \cos \vartheta \end{bmatrix} \tag{3.9}$$

with the precession parameters z, ϑ, ζ. These parameters are computed from the time series, cf. Nautical Almanac Office (1983), p. S19:

$$\begin{aligned} \zeta &= 2306\rlap{.}''2181\,T + 0\rlap{.}''30188\,T^2 + 0\rlap{.}''017998\,T^3 \\ z &= 2306\rlap{.}''2181\,T + 1\rlap{.}''09468\,T^2 + 0\rlap{.}''018203\,T^3 \\ \vartheta &= 2004\rlap{.}''3109\,T - 0\rlap{.}''42665\,T^2 - 0\rlap{.}''041833\,T^3\,. \end{aligned} \tag{3.10}$$

The parameter T represents the timespan expressed in Julian centuries of $36\,525$ mean solar days between the standard epoch J2000.0 and the epoch of observation. To give a numerical example consider an observation epoch J1990.5 which corresponds to $T = -0.095$. With T and Eq. (3.10) the numerical values $\zeta = -219\rlap{.}''0880$, $z = -219\rlap{.}''0809$, and $\vartheta = -190\rlap{.}''4134$ are obtained. Substitution of these values into Eq. (3.9) gives the following numerical precession matrix:

$$\underline{R}^P = \begin{bmatrix} 0.999997318 & 0.002124301 & 0.000923150 \\ -0.002124301 & 0.999997744 & -0.000000981 \\ -0.000923150 & -0.000000981 & 0.999999574 \end{bmatrix}.$$

Nutation. A graphic representation of nutation is given in Fig. 3.5. The mean vernal equinox at the observation epoch is denoted by E and the true equinox by E_t. The nutation matrix $\underline{R}^N$ is composed of three successive rotation matrices where both the nutation in longitude $\Delta\psi$ and the nutation in obliquity $\Delta\varepsilon$ can be treated as differential quantities:

$$\underline{R}^N = \underline{R}_1\{-(\varepsilon + \Delta\varepsilon)\}\,\underline{R}_3\{-\Delta\psi\}\,\underline{R}_1\{\varepsilon\}$$
$$= \begin{bmatrix} 1 & -\Delta\psi\cos\varepsilon & -\Delta\psi\sin\varepsilon \\ \Delta\psi\cos\varepsilon & 1 & -\Delta\varepsilon \\ \Delta\psi\sin\varepsilon & \Delta\varepsilon & 1 \end{bmatrix}. \tag{3.11}$$

The mean obliquity of the ecliptic ε has been determined, cf. Nautical Almanac Office (1983), p. S21, as

$$\varepsilon = 23°26'21.''448 - 46.''8150\,T - 0.''00059\,T^2 + 0.''001813\,T^3 \tag{3.12}$$

where T is the same time factor as in Eq. (3.10). The nutation parameters $\Delta\psi$ and $\Delta\varepsilon$ are computed from the harmonic series:

$$\Delta\psi = \sum_{i=1}^{106} a_i \sin(\sum_{j=1}^{5} e_j\,E_j) = -17.''2\,\sin\Omega_m + \dots$$
$$\Delta\varepsilon = \sum_{i=1}^{64} b_i \cos(\sum_{j=1}^{5} e_j\,E_j) = \quad 9.''2\,\cos\Omega_m + \dots \tag{3.13}$$

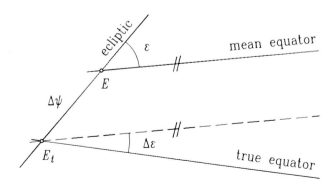

Fig. 3.5. Nutation

The amplitudes a_i, b_i as well as the integer coefficients e_j are tabulated for example in Nautical Almanac Office (1983), pp. S23–S26. The five fundamental arguments E_j describe mean motions in the sun-earth-moon system. The mean longitude Ω_m of moon's ascending node is one of the arguments. The moon's node retrogrades with a period of about 18.6 years and this period appears in the principal terms of the nutation series.

Sidereal time. The rotation matrix for sidereal time $\underline{R}^S$ is

$$\underline{R}^S = \underline{R}_3\{\Theta_0\}. \tag{3.14}$$

The computation of the apparent Greenwich sidereal time Θ_0 is shown in the section on time systems, cf. Sect. 3.3.

The WGS-84 system is defined by a uniform angular velocity ω_E, cf. Table 3.1. Consequently, instead of the apparent sidereal time the mean sidereal time must be used in the case of GPS for the rotation angle in Eq. (3.14).

Polar motion. To this point the instantaneous CEP has been obtained. The CEP must still be rotated into the CIO. This is achieved by means of the pole coordinates x_P, y_P which define the position of the CEP with respect to the CIO, cf. Fig. 3.6. The pole coordinates are determined by the International Earth Rotation Service (IERS) and are available upon request, cf. Feissel and McCarthy (1989). The rotation matrix for polar motion $\underline{R}^M$ is given by

$$\underline{R}^M = \underline{R}_2\{-x_P\}\,\underline{R}_1\{-y_P\} = \begin{bmatrix} 1 & 0 & x_P \\ 0 & 1 & -y_P \\ -x_P & y_P & 1 \end{bmatrix}. \tag{3.15}$$

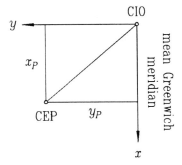

Fig. 3.6. Pole coordinates

The rotation matrices $\underline{R}^S$ and $\underline{R}^M$ are often combined to form a single matrix $\underline{R}^R$ for earth rotation:

$$\underline{R}^R = \underline{R}^M \underline{R}^S .$$ (3.16)

In the case of GPS the space-fixed coordinate system is already related to the CEP. Hence, $\underline{R}^R$ is the only rotation matrix which must be applied for the transformation into the terrestrial system.

3.3   Time systems

3.3.1   Definitions

Table 3.2 lists several time systems that are in current use.

Solar and sidereal times. A measure for earth rotation is the hour angle which is the angle between the meridian of a celestial body and a reference meridian (preferably the Greenwich meridian). Universal Time is defined by the Greenwich hour angle augmented by 12 hours of a fictitious sun uniformly orbiting in the equatorial plane. Sidereal time is defined by the hour angle of the vernal equinox. Taking the mean equinox as the reference leads to mean sidereal time and using the true equinox as a reference yields true or apparent sidereal time. Both, solar and sidereal times are not uniform since the angular velocity ω_E is not constant. The fluctuations are partly due to changes in the polar moment of inertia exerted by tidal deformation as well as other mass transports. Another reason has to do with the oscillations of the earth's rotational axis itself. In this case the universal time corrected for polar motion is denoted by UT1.

Table 3.2. Time systems

Periodic process	Time system
Earth rotation	Universal Time (UT)
	Greenwich Sidereal Time (Θ_0)
Earth revolution	Terrestrial Dynamic Time (TDT)
	Barycentric Dynamic Time (BDT)
Atomic oscillations	International Atomic Time (IAT)
	UT Coordinated (UTC)
	GPS Time (GPST)

Dynamic times. The time systems derived from planetary motions in the solar system are called dynamic times. The Barycentric Dynamic Time (BDT) is an inertial time system in the Newtonian sense and provides the time variable in the equations of motion. The quasi-inertial Terrestrial Dynamic Time (TDT) was formerly called ephemeris time and serves for the integration of the differential equations for the orbital motion of satellites around the earth.

Atomic times. A practical realization of the dynamic time system is achieved by the atomic time scales. GPS time also belongs to this system. The Universal Time Coordinated (UTC) system is a compromise. The unit of the system is the atomic second, but to keep the system close to UT and approximate civil time, integer leap seconds are inserted at distinct epochs. The GPS time system has a constant offset of 19 seconds with the International Atomic Time (IAT) and was coincident with UTC at the GPS standard epoch 1980, January $6.^d0$.

3.3.2 Conversions

The conversion between the times derived from earth rotation (i.e., the mean solar time corrected for polar motion UT1 and the apparent sidereal time Θ_0) is achieved by the formula

$$\Theta_0 = 1.002\,737\,9093 \text{ UT1} + \vartheta_0 + \Delta\psi \cos\varepsilon \,. \tag{3.17}$$

The first term in Eq. (3.17) accounts for the different scales of solar and sidereal times and the quantity ϑ_0 represents the actual sidereal time at Greenwich midnight (i.e., 0^h UT). The third term describes the projection of $\Delta\psi$ onto the equator and thus considers the effect of nutation. The mean sidereal time follows from Eq. (3.17) by neglecting the nutation term and is a part of the navigation message broadcast by the GPS satellites, cf. Sect. 4.4.2.

For ϑ_0 the following time series has been determined, cf. Nautical Almanac Office (1983), p. S13, by

$$\begin{aligned}
\vartheta_0 = {} & 24\,110\!\!\overset{s}{.}54841 + 8\,640\,184\!\!\overset{s}{.}812866\,T \\
& + 0\!\!\overset{s}{.}093104\,T^2 - 6\!\!\overset{s}{.}2 \cdot 10^{-6}\,T^3
\end{aligned} \tag{3.18}$$

where T is the same as in Eq. (3.10).

The time UT1 is related to UTC by the quantity dUT1 which is time dependent and is also reported by the IERS:

$$\text{UT1} = \text{UTC} + \text{dUT1} \,. \tag{3.19}$$

When the absolute value of dUT1 becomes larger than $0\overset{s}{.}9$, a leap second is inserted into the UTC system.

Instead of the dynamic time system itself, the atomic time system serves as reference in GPS. The following relations are defined:

$$\text{IAT} = \text{GPS} + 19\overset{s}{.}000 \qquad \text{constant offset}$$

$$\text{IAT} = \text{TDT} - 32\overset{s}{.}184 \qquad \text{constant offset} \tag{3.20}$$

$$\text{IAT} = \text{UTC} + 1\overset{s}{.}000\,n \qquad \begin{array}{l}\text{variable offset as leap sec-}\\ \text{onds are substituted.}\end{array}$$

The actual integer n is also reported by the IERS. In early 1992, for example, the integer value was $n = 26$ and thus GPS time differed by exactly seven seconds from UTC at this time.

3.3.3 Calendar

Definitions. The Julian Date (JD) defines the number of mean solar days elapsed since the epoch 4713 B.C., January $1\overset{d}{.}5$.

The Modified Julian Date (MJD) is obtained by subtracting $2\,400\,000.5$ days from JD. This convention saves digits and MJD commences at civil midnight instead of noon. For the sake of completeness, Table 3.3 with the Julian date for two standard epochs is given. This table enables, for example, the calculation of the parameter T for the GPS standard epoch. Subtracting the respective Julian dates and dividing by $36\,525$ (i.e., the number of days in a Julian century) yields $T = -0.199\,876\,7967$.

Date conversions. The relations for date conversions are taken from Montenbruck (1984) and are slightly modified so that they are only valid for an epoch between March 1900 and February 2100.

Let the civilian date be expressed by integer values for the year Y, month M, day D and a real value for the time in hours UT. Then

$$\text{JD} = \text{INT}[365.25\,y] + \text{INT}[30.6001\,(m+1)]$$
$$+ D + \text{UT}/24 + 1\,720\,981.5 \tag{3.21}$$

Table 3.3. Standard epochs

Civilian date	Julian date	Explanation
1980 January $6\overset{d}{.}0$	$2\,444\,244.5$	GPS standard epoch
2000 January $1\overset{d}{.}5$	$2\,451\,545.0$	Current standard epoch (J2000.0)

is the conversion into Julian date where INT denotes the integer part of a real number and y, m are given by

$$y = Y - 1 \quad \text{and} \quad m = M + 12 \qquad \text{if} \quad M \le 2$$
$$y = Y \qquad \text{and} \quad m = M \qquad \text{if} \quad M > 2.$$

The inverse transformation, that is the conversion of Julian date to civil date, is carried out stepwise. First, the auxiliary numbers

$$a = \text{INT}[JD + 0.5]$$
$$b = a + 1537$$
$$c = \text{INT}[(b - 122.1)/365.25]$$
$$d = \text{INT}[365.25\, c]$$
$$e = \text{INT}[(b - d)/30.6001]$$

are calculated. Afterwards, the civil date parameters are obtained from the relations

$$D = b - d - \text{INT}[30.6001\, e] + \text{FRAC}[JD + 0.5]$$
$$M = e - 1 - 12\,\text{INT}[e/14] \qquad (3.22)$$
$$Y = c - 4715 - \text{INT}[(7 + M)/10]$$

where FRAC denotes the fractional part of a number. As a by-product of date conversion, the day of week can be evaluated by the formula

$$N = \text{modulo}\{\text{INT}[JD + 0.5],\ 7\} \qquad (3.23)$$

where $N = 0$ denotes Monday, $N = 1$ means Tuesday, and so on. A further task is the calculation of the GPS week which is achieved by the relation

$$\text{WEEK} = \text{INT}[(JD - 2\,444\,244.5)/7]. \qquad (3.24)$$

The formulas given here can be used to proof the different dates in Table 3.3 or to verify the fact that the epoch J2000.0 corresponds to Saturday in the 1042nd GPS week.

4. Satellite orbits

4.1 Introduction

The applications of GPS depend substantially on knowing the satellite orbits. For single receiver positioning an orbital error is strongly correlated with the positional error. In relative positioning (according to a rule of thumb) relative baseline errors are equal to relative orbital errors.

Orbital information is either transmitted by the satellite as part of the broadcast message or can be obtained (typically many days after the observation) from several sources. Application of Selective Availability (SA) in the Block II satellites may lead to a degradation of the broadcast orbit in the 30 to 50 m range. Moreover, precise orbital information from the U.S. Defense Mapping Agency (DMA) for postprocessing will no longer be available to civilian users. The civil community, thus, must generate its own precise satellite ephemerides.

This chapter provides a review of orbital theory with emphasis on GPS orbits to acquaint the reader with the problems of computing ephemerides.

4.2 Orbit description

4.2.1 Keplerian motion

Orbital parameters. Assume two point masses m_1 and m_2 separated by the distance r. Considering for the moment only the attractive force between the masses and applying Newtonian mechanics, the movement of mass m_2 relative to m_1 is defined by a homogeneous differential equation of second order

$$\ddot{\underline{r}} + \frac{G\,(m_1 + m_2)}{r^3}\,\underline{r} = \underline{0} \tag{4.1}$$

where

$$\underline{r} \qquad \ldots \quad \text{relative position vector with } \|\underline{r}\| = r$$

$$\ddot{\underline{r}} = \frac{d^2\underline{r}}{dt^2} \quad \ldots \quad \text{relative acceleration vector}$$

$$G \qquad \ldots \quad \text{universal gravitational constant}$$

and the time parameter t being an inertial time realized by the GPS system time.

In the case of motion of an artificial earth satellite, in a first approximation, both bodies can be considered as point masses and the mass of the satellite can be neglected. The product of G and the earth's mass M_E is denoted as μ and is known as one defining parameter of the WGS-84 reference system, cf. Table (3.1), p. 27:

$$\mu = G\,M_E = 3\,986\,005 \cdot 10^8 \, \mathrm{m}^3 \cdot \mathrm{s}^{-2}\,.$$

The analytical solution of differential equation (4.1) can be found in textbooks on celestial mechanics, for example Brouwer and Clemence (1961) or Bucerius (1966), and leads to the well-known Keplerian motion defined by six orbital parameters which correspond to the six integration constants of the second-order vector equation (4.1). Satellite orbits may be restricted to elliptic motion, and the six associated parameters are listed in Table 4.1. The point of closest approach of the satellite with respect to the earth's center of mass is called perigee and the most distant position is the apogee. The intersection between the equatorial and the orbital plane with the unit sphere is termed the nodes, where the ascending node denotes the northward crossing of the equator. A graphical representation of the Keplerian orbit is given in Fig. 4.1.

The mean angular satellite velocity n (also known as the mean motion) with revolution period P follows from Kepler's Third Law which is given by

$$n = \frac{2\pi}{P} = \sqrt{\frac{\mu}{a^3}}\,. \tag{4.2}$$

For GPS orbits, the nominal semimajor axis is $a = 26\,560$ km. Substitution of a into Eq. (4.2) yields an orbital period of 12 sidereal hours. The ground track of the satellites thus repeats every sidereal day.

The instantaneous position of the satellite within its orbit is described by an angular quantity known (for historical reasons) as anomaly.

Table 4.1. Keplerian orbital parameters

Parameter	Notation
Ω	Right ascension of the ascending node
i	Inclination of orbital plane
ω	Argument of perigee
a	Semimajor axis of orbital ellipse
e	Numerical eccentricity of ellipse
T_0	Epoch of perigee passage

Table 4.2. Anomalies of the Keplerian orbit

Notation	Anomaly
$M(t)$	Mean anomaly
$E(t)$	Eccentric anomaly
$v(t)$	True anomaly

Table 4.2 lists commonly used anomalies. The mean anomaly $M(t)$ is a
mathematical abstraction while both the eccentric anomaly $E(t)$ and the
true anomaly $v(t)$ are geometrically produceable, cf. Fig. 4.1.

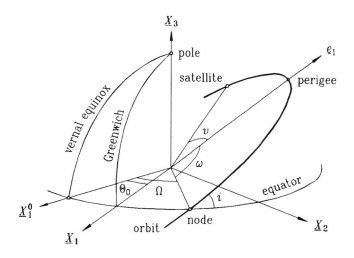

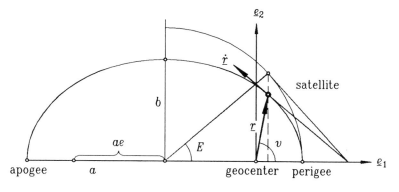

Fig. 4.1. Keplerian orbit

The three anomalies are related by the formulas

$$M(t) = n(t - T_0) \tag{4.3}$$

$$E(t) = M(t) + e \sin E(t) \tag{4.4}$$

$$v(t) = 2 \arctan \left[\sqrt{\frac{1+e}{1-e}} \tan \frac{E(t)}{2} \right] \tag{4.5}$$

where e denotes the eccentricity. Equation (4.3) holds by definition and shows that the mean anomaly can be used instead of T_0 as a defining parameter. Equation (4.4) is known as Kepler's equation and is obtained in the course of the analytical integration of Eq. (4.1). Finally, Eq. (4.5) follows from purely geometric relations as shown in the following paragraph.

To become more familiar with the various anomalies assume an orbit with a semidiurnal orbital period and an eccentricity of $e = 0.1$. At an epoch 3 hours after perigee passage the mean anomaly is $M = 90°.0000$. The calculation of the eccentric anomaly requires iteration and gives $E = 95°.7012$. The true anomaly is obtained as $v = 101°.3838$

Orbit representation. The coordinate system e_i, $i = 1,2$ which defines the orbital plane is shown in Fig. 4.1. The position vector r and the velocity vector $\dot{r} = dr/dt$ of the satellite can be represented by means of the eccentric as well as the true anomaly:

$$r = a \begin{bmatrix} \cos E - e \\ \sqrt{1 - e^2} \sin E \end{bmatrix} = r \begin{bmatrix} \cos v \\ \sin v \end{bmatrix} \tag{4.6}$$

$$r = a(1 - e \cos E) = \frac{a(1 - e^2)}{1 + e \cos v} \tag{4.7}$$

$$\dot{r} = \frac{n a^2}{r} \begin{bmatrix} -\sin E \\ \sqrt{1 - e^2} \cos E \end{bmatrix} = \sqrt{\frac{\mu}{a(1 - e^2)}} \begin{bmatrix} -\sin v \\ \cos v + e \end{bmatrix} \tag{4.8}$$

$$\dot{r} = \frac{n a^2}{r} \sqrt{1 - (e \cos E)^2} = \sqrt{\mu \left(\frac{2}{r} - \frac{1}{a} \right)}. \tag{4.9}$$

The components of the vector r are evident from the geometry in Fig. 4.1 where the semiminor axis b of the orbital ellipse is replaced by $a \sqrt{(1 - e^2)}$. The geocentric distance $r = r(E)$ corresponds to the norm $\|r(E)\|$ and follows from simple algebra. The representation $r = r(v)$ is known as polar equation of the ellipse, cf. e.g. Bronstein and Semendjajew (1969), p. 184.

Equation (4.5) can now be verified. For this purpose the first components of Eq. (4.6) are equated and Eq. (4.7) is substituted. This leads to $\cos v = (\cos E - e)/(1 - e \cos E)$. The identity $\tan(\alpha/2) = \sqrt{(1 - \cos\alpha)/(1 + \cos\alpha)}$ applied twice and using simple algebra yields Eq. (4.5).

The derivation of the velocity vector $\underline{\dot{r}} = \underline{\dot{r}}(E)$ requires knowledge of the time derivative $dE/dt = n\,a/r$. This result is obtained by differentiating Kepler's equation and substituting Eq. (4.7). The derivation of $\underline{\dot{r}} = \underline{\dot{r}}(v)$ is much more laborious and the result is given without proof. The norm of $\underline{\dot{r}}$ (i.e., the velocity $\dot{r}$) can be derived from either representation $\underline{\dot{r}}(E)$ or $\underline{\dot{r}}(v)$. The calculation of $\dot{r}(E)$ is straightforward; the proof of $\dot{r}(v)$ is left to the reader. It is worth noting that Eq. (4.9) when squared and divided by two relates kinetic energy on the left-hand side with potential energy on the right-hand side where a by definition is a constant. Hence, Eq. (4.9) can be recognized as the law of energy conservation in the earth-satellite system!

The transformation of $\underline{r}$ and $\underline{\dot{r}}$ into the equatorial system $\underline{X}_i^0$ is performed by a rotation matrix $\underline{R}$ and results in vectors denoted by $\underline{\varrho}$ and $\underline{\dot{\varrho}}$. The superscript "S" generally indicating a satellite is omitted here for simplicity. The vectors expressed in the orbital system must be considered as three-dimensional vectors for the transformation. Therefore, the axes $\underline{e}_1$, $\underline{e}_2$ are supplemented with an $\underline{e}_3$-axis which is orthogonal to the orbital plane. Since $\underline{r}$ and $\underline{\dot{r}}$ are vectors in the orbital plane (represented by $\underline{e}_1$, $\underline{e}_2$), their $\underline{e}_3$-component is zero.

The transformation is defined by

$$\underline{\varrho} = \underline{R}\,\underline{r}$$
$$\underline{\dot{\varrho}} = \underline{R}\,\underline{\dot{r}} \tag{4.10}$$

where the matrix $\underline{R}$ is composed of three successive rotation matrices, cf. Fig. 4.1, and is given by

$$\underline{R} = \underline{R}_3\{-\Omega\}\,\underline{R}_1\{-i\}\,\underline{R}_3\{-\omega\} =$$

$$=
\begin{bmatrix}
\begin{array}{c} \cos\Omega\cos\omega \\ -\sin\Omega\sin\omega\cos i \end{array} & \begin{array}{c} -\cos\Omega\sin\omega \\ -\sin\Omega\cos\omega\cos i \end{array} & \sin\Omega\sin i \\[2ex]
\begin{array}{c} \sin\Omega\cos\omega \\ +\cos\Omega\sin\omega\cos i \end{array} & \begin{array}{c} -\sin\Omega\sin\omega \\ +\cos\Omega\cos\omega\cos i \end{array} & -\cos\Omega\sin i \\[2ex]
\sin\omega\sin i & \cos\omega\sin i & \cos i
\end{bmatrix} \tag{4.11}$$

$$= \begin{bmatrix} \underline{e}_1, & \underline{e}_2, & \underline{e}_3 \end{bmatrix}.$$

The column vectors of the orthonormal matrix $\underline{R}$ are the axes of the orbital coordinate system represented in the equatorial system $\underline{X}_i^0$.

In order to rotate the system $\underline{X}_i^0$ into the earth-fixed system $\underline{X}_i$, an additional rotation through the angle Θ_0, the Greenwich sidereal time, is required. The transformation matrix, therefore, becomes

$$\underline{R}' = \underline{R}_3\{\Theta_0\}\,\underline{R}_3\{-\Omega\}\,\underline{R}_1\{-i\}\,\underline{R}_3\{-\omega\}\,. \tag{4.12}$$

The product $\underline{R}_3\{\Theta_0\}\,\underline{R}_3\{-\Omega\}$ can be expressed by a simple matrix $\underline{R}_3\{-\ell\}$ where $\ell = \Omega - \Theta_0$. Hence, Eq. (4.12) can be written in the form

$$\underline{R}' = \underline{R}_3\{-\ell\}\,\underline{R}_1\{-i\}\,\underline{R}_3\{-\omega\} \tag{4.13}$$

and matrix $\underline{R}'$ corresponds to matrix $\underline{R}$ if in Eq. (4.11) the parameter Ω is replaced by ℓ.

The inverse transformation to Eq. (4.10) aims to derive the Keplerian parameters from given position and velocity vectors, both expressed in an equatorial system such as $\underline{X}_i$. This problem is like the formulation of an initial value problem for solving the differential equation (4.1). The solution makes use of the fact that quantities like distances or angles are invariant with respect to rotation. Hence, the following equations are obtained:

$$\begin{aligned}
\|\underline{\varrho}\| &= \|\underline{r}\| \\
\|\underline{\dot{\varrho}}\| &= \|\underline{\dot{r}}\| \\
\|\underline{\varrho} \times \underline{\dot{\varrho}}\| &= \|\underline{r} \times \underline{\dot{r}}\| \\
\underline{\varrho} \cdot \underline{\dot{\varrho}} &= \underline{r} \cdot \underline{\dot{r}}\,.
\end{aligned} \tag{4.14}$$

In addition, by substituting Eqs. (4.6) and (4.8), the following equations can be derived:

$$\text{inner product:}\quad \underline{\varrho} \cdot \underline{\dot{\varrho}} = \sqrt{\mu a}\,(e\sin E) \tag{4.15}$$

$$\text{vector product:}\quad \|\underline{\varrho} \times \underline{\dot{\varrho}}\| = \sqrt{\mu a\,(1-e^2)}\,. \tag{4.16}$$

Note that the inner product is denoted by a dot, the cross or vector product by the symbol $\times$, and the norm of a vector by two $\|$ symbols.

The inverse transformation can be solved as follows. First, from the given $\underline{\varrho}$ and $\underline{\dot{\varrho}}$ the geocentric distance r and the velocity $\dot{r}$ are calculated. With these two quantities the semimajor axis a follows from Eq. (4.9). With a and r determined, $e\cos E$ can be established using Eq. (4.7) and $e\sin E$ can be established using Eq. (4.15). Hence, the eccentricity e and the eccentric anomaly E, and consequently the mean and true anomalies M and v can be calculated. The vector product of $\underline{\varrho}$ and $\underline{\dot{\varrho}}$ is equivalent to the vector $\underline{c}$ of angular momentum and is directed orthogonal to the orbital plane. Therefore, the vector $\underline{c}$ after normalisation is identical to the vector $\underline{e}_3$ in Eq. (4.11) from which the parameters i and ℓ can be deduced. According to Eq. (4.16) the norm of the vector $\underline{c}$ allows for a check of the previously calculated parameters a and e. For the determination of ω, the unit vector $\underline{k} = (\cos \ell, \sin \ell, 0)^T$ directed to the ascending node is defined where "T" denotes transposition. From Fig. 4.1 one can get the relations $\underline{\varrho} \cdot \underline{k} = r \cos (\omega + v)$ and $\underline{\varrho} \cdot \underline{X}_3 = r \sin i \sin (\omega + v)$. And, since r, v, i are already known, the two equations can be uniquely solved for ω.

For a numerical example assume a satellite orbiting in a Kepler ellipse with the following parameters: $a = 26\,000$ km, $e = 0.1$, $\omega = -140°$, $i = 60°$, $\ell = 110°$. To calculate the position and velocity vector in the earth-fixed equatorial system at an epoch where the eccentric anomaly is $E = 45°$, the vectors are first calculated in the orbital plane using Eqs. (4.6) through (4.9). Then the transformation into the equatorial system is performed by means of Eq. (4.10) using the rotation matrix $\underline{R}'$. The final result is

$$\underline{\varrho} = (11\,465,\ 3\,818,\ -20\,923)^T \qquad [\text{km}]$$

$$\dot{\underline{\varrho}} = (-1.2651,\ 3.9960,\ -0.3081)^T \quad [\text{km·s}^{-1}].$$

It is recommended for the reader also to perform the inverse transformation where the position vector $\underline{\varrho}$ and the velocity vector $\dot{\underline{\varrho}}$ are given and the Keplerian parameters are determined.

In addition to the fixed orbital system $\underline{e}_i$, another orthonormal system $\underline{e}_i^*$ may be defined. This system rotates around the $\underline{e}_3$-axis because the $\underline{e}_1^*$-axis always points towards the instantaneous satellite position. Hence, the unit vectors $\underline{e}_i^*$ can be derived from the position and velocity vectors by

$$\underline{e}_1^* = \frac{\underline{\varrho}}{\|\underline{\varrho}\|} \qquad \underline{e}_3^* = \frac{\underline{\varrho} \times \dot{\underline{\varrho}}}{\|\underline{\varrho} \times \dot{\underline{\varrho}}\|} = \underline{e}_3 \qquad \underline{e}_2^* = \underline{e}_3 \times \underline{e}_1 . \qquad (4.17)$$

Note that the base vectors $\underline{e}_i^*$ correspond to the column vectors of a modified rotation matrix $\underline{R}^*$ in Eq. (4.11) where the parameter ω is replaced by $(\omega + v)$.

The transformation of a change $d\underline{\varrho}$ in the position vector into the orbital system $\underline{e}_i^*$ results in a vector $\underline{e} = \{e_i\}$ with its components reckoned along the respective axes $\underline{e}_i^*$:

$$e_1 = \underline{e}_1^* \cdot d\underline{\varrho} \qquad \text{radial or across-track component}$$

$$e_2 = \underline{e}_2^* \cdot d\underline{\varrho} \qquad \text{along-track component} \qquad\qquad (4.18)$$

$$e_3 = \underline{e}_3^* \cdot d\underline{\varrho} \qquad \text{out-of-plane component}.$$

Inversely, $d\underline{\varrho}$ is calculated if the vector $\underline{e}$ is given. The solution follows from the inversion of Eq. (4.18) and leads to

$$d\underline{\varrho} = \underline{R}^* \underline{e} . \qquad\qquad (4.19)$$

For a numerical examination assume a change $d\underline{\varrho} = (0.1, 1.0, -0.5)^T$ [km] in the satellite's position of the previous example. Applying Eqs. (4.17) and (4.18) gives $\underline{e} = (0.638, 0.914, 0.128)^T$ [km].

Differential relations. The derivatives of $\underline{\varrho}$ and $\underline{\dot{\varrho}}$ with respect to the six Keplerian parameters are required in one of the subsequent sections. The differentiation can be separated into two groups because in Eq. (4.10) the vectors $\underline{r}$ and $\underline{\dot{r}}$ depend only on the parameters a, e, T_0 whereas the matrix $\underline{R}$ is only a function of the remaining parameters ω, i, Ω.

In order to obtain the derivatives of $\underline{r}$ and $\underline{\dot{r}}$, it is advantageous to differentiate the vectors represented by the eccentric anomaly. As a first step, the derivatives of the quantities n, E, r with respect to the parameters a, e, T_0 are calculated. By differentiating Eq. (4.2) one gets

$$\frac{\partial n}{\partial a} = -\frac{3}{2}\frac{n}{a} \qquad \frac{\partial n}{\partial e} = 0 \qquad \frac{\partial n}{\partial T_0} = 0 \,. \tag{4.20}$$

Differentiating the Kepler equation (4.4) and substituting Eq. (4.20) leads to

$$\frac{\partial E}{\partial a} = -\frac{3}{2}\frac{M}{r} \qquad \frac{\partial E}{\partial e} = \frac{a}{r}\sin E \qquad \frac{\partial E}{\partial T_0} = -n\frac{a}{r} \,. \tag{4.21}$$

Finally, the differentiation of Eq. (4.7) yields

$$\frac{\partial r}{\partial a} = (1 - e\cos E) + a\,e\,\sin E\,\frac{\partial E}{\partial a}$$

$$\frac{\partial r}{\partial e} = -a\,\cos E + a\,e\,\sin E\,\frac{\partial E}{\partial e} \tag{4.22}$$

$$\frac{\partial r}{\partial T_0} = a\,e\,\sin E\,\frac{\partial E}{\partial T_0} \,.$$

Now, the desired result follows after applying the chain rule to Eqs. (4.6) and (4.8), and after algebraic transformations where $dm = -n\,dT_0$ is substituted:

$$\frac{\partial \underline{r}}{\partial a} = \left[\begin{array}{c} \cos E - e + \dfrac{3}{2}\dfrac{a}{r}\,M\sin E \\[2mm] \sqrt{1 - e^2}\,(\sin E - \dfrac{3}{2}\dfrac{a}{r}\,M\cos E) \end{array} \right] \tag{4.23}$$

$$\frac{\partial \underline{r}}{\partial e} = \left[\begin{array}{c} -a\,(1 + \dfrac{a}{r}\,\sin^2 E) \\[2mm] \dfrac{a^2\sin E}{r\,\sqrt{1 - e^2}}\,(\cos E - e) \end{array} \right] \tag{4.24}$$

$$\frac{\partial \underline{r}}{\partial m} = \left[\begin{array}{c} -\dfrac{a^2}{r} \sin E \\[2mm] \dfrac{a^2}{r} \sqrt{1-e^2} \cos E \end{array} \right] \tag{4.25}$$

$$\frac{\partial \underline{\dot{r}}}{\partial a} = \frac{an}{2r} \left[\begin{array}{c} \sin E - \dfrac{3a}{r} M \left(\dfrac{ae}{r} \sin^2 E - \cos E \right) \\[3mm] \sqrt{1-e^2} \left(\dfrac{3a}{r} M \sin E \left(\dfrac{ae}{r} \cos E + 1 \right) - \cos E \right) \end{array} \right] \tag{4.26}$$

$$\frac{\partial \underline{\dot{r}}}{\partial e} = \frac{a^2 n}{r} \left[\begin{array}{c} \dfrac{a^2}{2r^2} \left(2e \sin E + \sin(2E)(e \cos E - 2) \right) \\[3mm] -\dfrac{e}{\sqrt{1-e^2}} \cos E + \dfrac{a}{r} \sqrt{1-e^2} \left(\dfrac{a}{r} \sin^2 E - \cos^2 E \right) \end{array} \right] \tag{4.27}$$

$$\frac{\partial \underline{\dot{r}}}{\partial m} = -\frac{na^4}{r^3} \left[\begin{array}{c} \cos E - e \\[2mm] \sqrt{1-e^2} \sin E \end{array} \right] . \tag{4.28}$$

Considering Eq. (4.11), the differentiation of the matrix $\underline{R}$ with respect to the parameters ω, i, Ω is simple and does not pose any problem.

Hence, the differential relations are given by

$$\begin{aligned} d\underline{\varrho} &= \underline{R} \, \frac{\partial \underline{r}}{\partial a} \, da + \underline{R} \, \frac{\partial \underline{r}}{\partial e} \, de + \underline{R} \, \frac{\partial \underline{r}}{\partial m} \, dm \\[2mm] &+ \frac{\partial \underline{R}}{\partial \omega} \, \underline{r} \, d\omega + \frac{\partial \underline{R}}{\partial i} \, \underline{r} \, di + \frac{\partial \underline{R}}{\partial \Omega} \, \underline{r} \, d\Omega \\[4mm] d\underline{\dot{\varrho}} &= \underline{R} \, \frac{\partial \underline{\dot{r}}}{\partial a} \, da + \underline{R} \, \frac{\partial \underline{\dot{r}}}{\partial e} \, de + \underline{R} \, \frac{\partial \underline{\dot{r}}}{\partial m} \, dm \\[2mm] &+ \frac{\partial \underline{R}}{\partial \omega} \, \underline{\dot{r}} \, d\omega + \frac{\partial \underline{R}}{\partial i} \, \underline{\dot{r}} \, di + \frac{\partial \underline{R}}{\partial \Omega} \, \underline{\dot{r}} \, d\Omega . \end{aligned} \tag{4.29}$$

In Eqs. (4.23) through (4.28) the mean or eccentric anomaly is contained. Hence, these derivatives with respect to the parameters a, e, m are time dependent. Because of the appearance of $\underline{r}$ and $\underline{\dot{r}}$, all terms in Eq. (4.29) are time dependent although the derivatives of the matrix $\underline{R}$ with respect to the parameters ω, i, Ω are constant.

4.2.2 Perturbed motion

The Keplerian orbit is a theoretical orbit and does not include actual perturbations. Consequently, disturbing accelerations $d\ddot{\varrho}$ must be added to
Eq. (4.1) which is now expressed in the equatorial system. The perturbed
motion thus is based on an inhomogeneous differential equation of the second
order

$$\ddot{\varrho} + \frac{\mu}{\varrho^3}\, \varrho = d\ddot{\varrho}\,. \tag{4.30}$$

One should note that, for GPS satellites, the acceleration $\|\ddot{\varrho}\|$ due to the
central attractive force μ/ϱ^2 is at least 10^4 times larger than the disturbing
accelerations. Hence, for the analytical solution of Eq. (4.30), perturbation theory can be applied where, initially, only the homogeneous part of
the equation is considered. This leads to a Keplerian orbit defined by the
six parameters p_{i0}, $i = 1,\ldots,6$ at the reference epoch t_0. Each disturbing
acceleration $d\ddot{\varrho}$ causes temporal variations $\dot{p}_{i0} = dp_{i0}/dt$ in the orbital parameters. Therefore, at an arbitrary epoch t, the parameters p_i describing
the so-called osculating ellipse are given by

$$p_i = p_{i0} + \dot{p}_{i0}\,(t - t_0)\,. \tag{4.31}$$

In order to get time derivatives $\dot{p}_{i0}$, the Keplerian motion is compared to the
perturbed motion. In the first case, the parameters p_i are constant whereas
in the second case they are time dependent. Thus, for the position and
velocity vector of the perturbed motion one may write

$$\begin{aligned}
\varrho &= \varrho\{t, p_i(t)\} \\
\dot{\varrho} &= \dot{\varrho}\{t, p_i(t)\}\,.
\end{aligned} \tag{4.32}$$

Differentiating the above equations with respect to time and taking into
account Eq. (4.30) leads to

$$\dot{\varrho} = \frac{\partial \varrho}{\partial t} + \sum_{i=1}^{6} \left(\frac{\partial \varrho}{\partial p_i} \frac{dp_i}{dt} \right) \tag{4.33}$$

$$\ddot{\varrho} = \frac{\partial \dot{\varrho}}{\partial t} + \sum_{i=1}^{6} \left(\frac{\partial \dot{\varrho}}{\partial p_i} \frac{dp_i}{dt} \right) = -\frac{\mu}{\varrho^3}\, \varrho + d\ddot{\varrho}\,. \tag{4.34}$$

Since for any epoch t an (osculating) ellipse is defined, the Eqs. (4.33) and
(4.34) must also hold for a Keplerian motion. Evidently, equivalence is

obtained with the following conditions

$$\sum_{i=1}^{6} \left(\frac{\partial \underline{\varrho}}{\partial p_i} \frac{dp_i}{dt} \right) = \underline{0}$$

$$\sum_{i=1}^{6} \left(\frac{\partial \underline{\dot{\varrho}}}{\partial p_i} \frac{dp_i}{dt} \right) = d\underline{\ddot{\varrho}}.$$

(4.35)

In the following, for simplicity, only one disturbing acceleration is considered. The two vector equations (4.35) correspond to six linear equations which, in vector notation, are given by

$$\underline{A}\,\underline{u} = \underline{l}$$

(4.36)

where

$$\underline{A} = \begin{bmatrix} \dfrac{\partial \underline{\varrho}}{\partial a} & \dfrac{\partial \underline{\varrho}}{\partial e} & \dfrac{\partial \underline{\varrho}}{\partial m} & \dfrac{\partial \underline{\varrho}}{\partial \omega} & \dfrac{\partial \underline{\varrho}}{\partial i} & \dfrac{\partial \underline{\varrho}}{\partial \Omega} \\[2mm] \dfrac{\partial \underline{\dot{\varrho}}}{\partial a} & \dfrac{\partial \underline{\dot{\varrho}}}{\partial e} & \dfrac{\partial \underline{\dot{\varrho}}}{\partial m} & \dfrac{\partial \underline{\dot{\varrho}}}{\partial \omega} & \dfrac{\partial \underline{\dot{\varrho}}}{\partial i} & \dfrac{\partial \underline{\dot{\varrho}}}{\partial \Omega} \end{bmatrix}$$

$$= \begin{bmatrix} R\dfrac{\partial \underline{r}}{\partial a} & R\dfrac{\partial \underline{r}}{\partial e} & R\dfrac{\partial \underline{r}}{\partial m} & \underline{r}\dfrac{\partial R}{\partial \omega} & \underline{r}\dfrac{\partial R}{\partial i} & \underline{r}\dfrac{\partial R}{\partial \Omega} \\[2mm] R\dfrac{\partial \underline{\dot{r}}}{\partial a} & R\dfrac{\partial \underline{\dot{r}}}{\partial e} & R\dfrac{\partial \underline{\dot{r}}}{\partial m} & \underline{\dot{r}}\dfrac{\partial R}{\partial \omega} & \underline{\dot{r}}\dfrac{\partial R}{\partial i} & \underline{\dot{r}}\dfrac{\partial R}{\partial \Omega} \end{bmatrix}$$

$$\underline{u} = \begin{bmatrix} \dfrac{da}{dt} & \dfrac{de}{dt} & \dfrac{dm}{dt} & \dfrac{d\omega}{dt} & \dfrac{di}{dt} & \dfrac{d\Omega}{dt} \end{bmatrix}^T$$

$$= \begin{bmatrix} \dot{a} & \dot{e} & \dot{m} & \dot{\omega} & \dot{i} & \dot{\Omega} \end{bmatrix}^T$$

$$\underline{l} = \begin{bmatrix} \underline{0} \\ d\underline{\ddot{\varrho}} \end{bmatrix}.$$

The 6×6 matrix $\underline{A}$ requires the derivatives of $\underline{\varrho}$ and $\underline{\dot{\varrho}}$ with respect to the Keplerian parameters which have been developed in the preceding section, cf. Eq. (4.29). The 6×1 vector $\underline{l}$ contains the disturbing acceleration. Finally, the six unknown time derivatives appear in the 6×1 vector $\underline{u}$. The inversion of the system (4.36) leads to Lagrange's equations where the disturbing potential R, associated with the disturbing acceleration by $d\underline{\ddot{\varrho}} = \mathrm{grad}\,R$, has been introduced, cf. for example Seeber (1989), Eq. (3.98):

$$\dot{a} = \frac{2}{n\,a}\frac{\partial R}{\partial m}$$

$$\dot{e} = \frac{1-e^2}{n\,a^2\,e}\frac{\partial R}{\partial m} - \frac{\sqrt{1-e^2}}{n\,a^2\,e}\frac{\partial R}{\partial \omega}$$

$$\dot{m} = -\frac{2}{n\,a}\frac{\partial R}{\partial a} - \frac{1-e^2}{n\,a^2\,e}\frac{\partial R}{\partial e}$$

$$\dot{\omega} = \frac{\sqrt{1-e^2}}{n\,a^2\,e}\frac{\partial R}{\partial e} - \frac{\cos i}{n\,a^2\,\sqrt{1-e^2}\,\sin i}\frac{\partial R}{\partial i} \qquad (4.37)$$

$$\dot{i} = \frac{\cos i}{n\,a^2\,\sqrt{1-e^2}\,\sin i}\frac{\partial R}{\partial \omega} - \frac{1}{n\,a^2\,\sqrt{1-e^2}\,\sin i}\frac{\partial R}{\partial \Omega}$$

$$\dot{\Omega} = \frac{1}{n\,a^2\,\sqrt{1-e^2}\,\sin i}\frac{\partial R}{\partial i}\,.$$

Note that the system (4.37) fails for $e = 0$ or $i = 0$. This singularity can be avoided by the substitution of auxiliary parameters, cf. Bucerius (1966), p. 193 or Arnold (1970), p. 28.

The Lagrange equations presuppose that the disturbing potential R is expressed in function of the Keplerian parameters. When the acceleration $d\underline{\ddot{\varrho}}$ is represented by components K_i along the axes $\underline{e}_i^*$, the system (4.37) can be transformed using the identity

$$\frac{\partial R}{\partial p_i} = \mathrm{grad}\,R \cdot \frac{\partial \underline{\varrho}}{\partial p_i} = (K_1\,\underline{e}_1^* + K_2\,\underline{e}_2^* + K_3\,\underline{e}_3^*) \cdot \frac{\partial \underline{\varrho}}{\partial p_i}\,. \qquad (4.38)$$

The simple but cumbersome algebra leads to the Gaussian equations. The result is taken from Seeber (1989), Eq. (3.101) where K_1 and K_3 are interchanged and slight modifications are introduced:

$$\dot{a} = \frac{2}{n\,\sqrt{1-e^2}}\,[e\,\sin v\,K_1 + (1 + e\,\cos v)\,K_2]$$

$$\dot{e} = \frac{\sqrt{1-e^2}}{n\,a}\,[\sin v\,K_1 + (\cos E + \cos v)\,K_2]$$

$$\dot{m} = \frac{1-e^2}{n\,a\,e}\left[\left(\frac{-2e}{1+e\,\cos v} + \cos v\right)K_1\right.$$
$$\left. - \left(1 + \frac{1}{1+e\,\cos v}\right)\sin v\,K_2\right]$$

(continued on next page)

$$\dot{\omega} = \frac{\sqrt{1-e^2}}{n\,a\,e} \left[-\cos v\, K_1 + \left(1 + \frac{1}{1 + e\cos v}\right) \sin v\, K_2 \right.$$
$$\left. - \frac{e\,\sin(\omega + v)}{1 + e\cos v} \cot i\, K_3 \right]$$

$$i = \frac{r\,\cos(\omega + v)}{n\,a^2\,\sqrt{1-e^2}} K_3$$

$$\dot{\Omega} = \frac{r\,\sin(\omega + v)}{n\,a^2\,\sqrt{1-e^2}\,\sin i} K_3 \,.$$

$$(4.39)$$

Note that in the temporal variations i and $\dot{\Omega}$ only the component orthogonal to the orbital plane, K_3, appears whereas the variations of $\dot{a}$, $\dot{e}$, $\dot{m}$ are affected by both components in the orbital plane, K_1, K_2.

4.2.3 Disturbing accelerations

In reality many disturbing accelerations act on a satellite and are responsible for the temporal variations of the Keplerian elements. Roughly speaking, they can be divided into two groups, namely those of gravitational and non-gravitational origin, see Table 4.3. Because the GPS satellites are orbiting at an altitude of approximately 20 000 km, the indirect effects as well as air drag may be neglected. On the other hand, the shape (and thus the cross section) of the satellites is irregular which renders the modeling of solar radiation more difficult. The variety of materials used for the satellites each has a different heat-absorption which results in additional and complicated perturbing accelerations. Also, accelerations may arise from leaks in the container for the gas-propellant as mentioned by Lichten and Neilan (1990).

    To get an idea of the effect of disturbing accelerations, an example is computed by assuming a constant disturbance $d\ddot{\varrho} = 10^{-9}\,\mathrm{m \cdot s^{-2}}$ acting on a

Table 4.3. Sources for disturbing accelerations

Gravitational	Nonsphericity of the earth,
	Tidal attraction (direct and indirect)
Nongravitational	Solar radiation pressure (direct and indirect),
	Air drag,
	Relativistic effects,
	Others (solar wind, magnetic field forces, etc.)

GPS satellite. The associated shift in the position of the satellite results from double integration over time t and yields $d\varrho = d\ddot{\varrho}\,(t^2/2)$. Substituting the numerical value $t \approx 12$ hours gives the shift after one revolution which is $d\varrho \approx 1\,\mathrm{m}$. This value can be considered as typical.

Nonsphericity of the earth. The earth's potential V can be represented by a spherical harmonic expansion, cf. Heiskanen and Moritz (1967), p. 342 by

$$V = \frac{\mu}{r}\left[1 - \sum_{n=2}^{\infty} \left(\frac{a_E}{r}\right)^n J_n\, P_n(\sin\varphi)\right.$$

$$\left. - \sum_{n=2}^{\infty}\sum_{m=1}^{n} \left(\frac{a_E}{r}\right)^n \left[J_{nm}\cos m\lambda + K_{nm}\sin m\lambda\right] P_{nm}(\sin\varphi)\right]$$

$$(4.40)$$

where a_E is the semimajor axis of the earth, r is the geocentric distance of the satellite, and φ, λ are its latitude and longitude. The J_n, J_{nm}, K_{nm} denote the zonal and tesseral coefficients of the harmonic development known from an earth model. Finally, P_n are the Legendre polynomials and P_{nm} are the associated Legendre functions.

The first term on the right-hand side of Eq. (4.40), μ/r, represents the potential V_0 for a spherical earth, and its gradient, $\mathrm{grad}(\mu/r) = (\mu/r^3)\,\underline{r}$, was considered as the central force for the Keplerian motion. Hence, the disturbing potential R is given by the difference

$$R = V - V_0 \,. \qquad\qquad (4.41)$$

It is shown below that the disturbing acceleration due to J_2, the term representing the oblateness, is smaller by a factor of 10^4 than the acceleration due to V_0. On the other hand, the oblateness term is approximately three orders of magnitude larger than any other coefficient. Rizos and Stolz (1985) pointed out that a subset of the coefficients complete up to degree and order eight is sufficient for GPS satellite arcs of a few revolutions.

A numerical assessment of the central acceleration of GPS satellites gives $\|\ddot{\underline{r}}\| = (\mu/r^2) \approx 0.57\,\mathrm{m\cdot s^{-2}}$. The acceleration corresponding to the disturbing potential R is given by $\|d\ddot{\underline{r}}\| \approx \|\partial R/\partial r\| = 3\mu\,(a_E/r^2)^2 J_2\, P_2(\sin\varphi)$. The latitude of a satellite can only reach $\varphi = 55°$, the value of its orbital inclination. The maximum of the function $P_2(\sin\varphi) = \frac{1}{2}\,(3\sin\varphi^2 - 1)$ becomes therefore 0.5. Finally, with $J_2 \approx 1.1\cdot 10^{-3}$, cf. Table 3.1, the numerical value $\|d\ddot{\underline{r}}\| \approx 5\cdot 10^{-5}\,\mathrm{m\cdot s^{-2}}$ is obtained.

Tidal effects. Consider a celestial body with pointmass m_b and the geocentric position vector $\underline{\varrho}_b$, see Fig. 4.2. Note that the geocentric angle z between

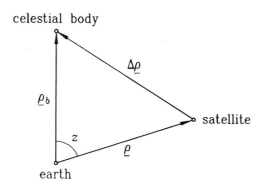

Fig. 4.2. Three-body problem

the celestial body and the satellite can be expressed as a function of ϱ_b and ϱ, the latter denoting the geocentric position vector of the satellite, by

$$\cos z = \frac{\varrho_b}{\|\varrho_b\|} \cdot \frac{\varrho}{\|\varrho\|}. \tag{4.42}$$

The additional mass exerts an acceleration with respect to the earth as well as with respect to the satellite. For the perturbed motion of the satellite around the earth, only the difference of the two accelerations is relevant; consequently, the disturbing acceleration is given by

$$d\ddot{\varrho} = G\,m_b \left[\frac{\varrho_b - \varrho}{\|\varrho_b - \varrho\|^3} - \frac{\varrho_b}{\|\varrho_b\|^3} \right]. \tag{4.43}$$

Among all the celestial bodies in the solar system, only the sun and the moon must be considered because the effects of the planets are negligible. The geocentric position vector of the sun and the moon are obtained by evaluating known analytical expressions for their motion.

The maximum of the perturbing acceleration is reached when the three bodies in Fig. 4.2 are situated in a straight line. In this case Eq. (4.43) reduces to $\|d\ddot{\varrho}\| = G\,m_b\,(1/\|\varrho_b - \varrho\|^2 - 1/\|\varrho_b\|^2)$. For a numerical assessment the corresponding numerical values for the sun ($G\,m_b \approx 1.3 \cdot 10^{20}\,\mathrm{m^3 \cdot s^{-2}}$, $\varrho_b \approx 1.5 \cdot 10^{11}\,\mathrm{m}$) and of the moon ($G\,m_b \approx 4.9 \cdot 10^{12}\,\mathrm{m^3 \cdot s^{-2}}$, $\varrho_b \approx 3.8 \cdot 10^8\,\mathrm{m}$) are substituted. The resulting numerical values for the perturbing acceleration are $2 \cdot 10^{-6}\,\mathrm{m \cdot s^{-2}}$ for the sun and $5 \cdot 10^{-6}\,\mathrm{m \cdot s^{-2}}$ for the moon.

Apart from the direct effect of the tide generating bodies, indirect effects due to the tidal deformation of the solid earth and the oceanic tides must be taken into account. Considering only the tidal potential W_2 of second degree, the disturbing potential R due to the tidal deformation of the solid earth, cf. Melchior (1978), is given by

$$R = k \left(\frac{a_E}{\varrho}\right)^3 W_2 = \frac{1}{2} k\, G\, m_b \frac{a_E{}^5}{(\varrho\, \varrho_b)^3} (3\cos^2 z - 1) \qquad (4.44)$$

with $k \approx 0.3$ being one of the Love numbers. The associated acceleration of the satellite is in the order of 10^{-9} m·s^{-2} as the reader may verify.

The indirect effect due to the oceanic tides is more difficult to model. Tidal charts with the distribution of the oceanic tides are required. Such charts have been published by Schwiderski (1981). In addition, loading coefficients as computed by Farrell (1972) are needed. These coefficients describe the response of the solid earth to the load of the oceanic water masses. The perturbing acceleration is again in the order of 10^{-9} m·s^{-2}.

As a consequence of the tidal deformation and the oceanic loading, the geocentric position vector $\underline{\varrho}_R$ of an observing site varies with time. This variation must be taken into account when modeling receiver dependent biases in the observation equations, cf. for example McCarthy (1989).

Solar radiation pressure. Following Fliegel et al. (1985), the perturbing acceleration due to the direct solar radiation pressure has two components. The principal component $d\underline{\ddot{\varrho}}_1$ is directed away from the sun and the smaller component $d\underline{\ddot{\varrho}}_2$ acts along the satellite's y-axis. This is an axis orthogonal to both the vector pointing to the sun and the antenna which is nominally directed towards the earth-center.

The principal component is usually modeled by

$$d\underline{\ddot{\varrho}}_1 = \nu\, K\, \varrho_S^2 \frac{\underline{\varrho} - \underline{\varrho}_S}{\|\underline{\varrho} - \underline{\varrho}_S\|^3} \qquad (4.45)$$

where $\underline{\varrho}_S$ denotes the geocentric position vector of the sun. The factor K depends linearly on the solar radiation constant P_S, a factor C_r defining the reflective properties of the satellite, and the area-to-mass ratio of the satellite. The quantity ν is an eclipse factor which is zero when the satellite is in the earth's shadow. This occurs twice a year for each satellite when the sun is in or near the orbital plane. Such an eclipse lasts approximately one hour. The eclipse factor equals one when the satellite is in sunlight and for the penumbra regions the relation $0 < \nu < 1$ holds.

The magnitude of $d\underline{\ddot{\varrho}}_1$ is in the order of 10^{-7} m·s^{-2}. Hence, an accurate model for the factors K and ν is required even for short arcs. The modeling is extremely difficult since the factor P_S varies unpredictably over the year and a single factor C_r is not adequate for the satellite. Although the mass in orbit is usually well-known, the irregular shape of the satellites does not allow for an exact determination of the area-to-mass-ratio. A further problem is the modeling of the earth's penumbra and the assignment of an eclipse factor, particularly in the transition zone between illumination and shadow. Some models are given in Landau (1988).

The component $d\ddot{\underline{\varrho}}_2$ is often called y-bias and is believed to be caused by a combination of misalignments of the solar panels and thermal radiation along the y-axis. Since the magnitude of this bias can remain constant for several weeks, it is usually introduced as an unknown parameter which is determined in the course of the orbit determination. Note that this bias is two orders of magnitudes smaller than the principal term, cf. Feltens (1988).

That portion of the solar radiation pressure which is reflected back from the earth's surface causes an effect called albedo. In the case of GPS, the associated perturbing accelerations are smaller than the y-bias and can be neglected.

Relativistic effects. The relativistic effect on the satellite orbit is caused by the earth's gravity field and gives rise to a perturbing acceleration which, simplified, is given, cf. Beutler (1991), Eq. (2.5), by

$$d\ddot{\underline{\varrho}} = -\frac{3\mu^2\,a\,(1-e^2)}{c^2}\,\frac{\underline{\varrho}}{\varrho^5} \tag{4.46}$$

where c denotes the velocity of light. Numerically assessed, the perturbing acceleration results in an order of $3\cdot10^{-10}\,\mathrm{m\cdot s^{-2}}$, cf. Zhu and Groten (1988). This effect is smaller than the indirect effects by one order of magnitude and is only mentioned for the sake of completeness.

4.3 Orbit determination

Here, orbit determination essentially means the determination of orbital parameters and satellite clock biases. In principle, the problem is inverse to the navigational or surveying goal. In the fundamental equation for the range ϱ or the range rate $\dot{\varrho}$ between the observing site R and the GPS satellite S,

$$\varrho = \|\underline{\varrho}^S - \underline{\varrho}_R\| \tag{4.47}$$

$$\dot{\varrho} = \frac{\underline{\varrho}^S - \underline{\varrho}_R}{\|\underline{\varrho}^S - \underline{\varrho}_R\|} \cdot \dot{\underline{\varrho}}^S, \tag{4.48}$$

the position vector $\underline{\varrho}^S$ and the velocity vector $\dot{\underline{\varrho}}^S$ of the satellite are considered unknown whereas the coordinates of the observing site are assumed to be known in a geocentric system. Note that the satellite vectors are indicated with a superscript.

From the observation equations one may state that the position vector $\underline{\varrho}^S$ is a function of ranges, whereas the velocity vector $\underline{\dot{\varrho}}^S$ is determined by range rates. The ranges in Eq. (4.47) are obtained with high precision as outlined in Sect. 6.1. This is particularly true for delta range data since biases are eliminated by differencing the ranges. The range rates in Eq. (4.48) are less accurate and are derived from frequency shifts due to the Doppler effect. At present, the observations for the orbit determination are performed at terrestrial points, but in the near future GPS data will also be gained from orbiting receivers such as the TOPEX/Poseidon mission, cf. Lichten and Neilan (1990).

In the following, the satellite clock biases and other parameters are neglected. The actual orbit determination which is performed in two steps is emphasized. First, a Kepler ellipse is fitted to the observations. In the second step, this ellipse serves as reference for the subsequent improvement of the orbit by taking into account perturbing accelerations.

4.3.1   Keplerian orbit

For the moment, it is assumed that both the position and the velocity vector of the satellite have been derived from observations. Now, the question arises of how to use these data for the derivation of the Keplerian parameters.

The position and velocity vector given at the same epoch t define an initial value problem which has been treated in Sect. 4.2.1. Recall that the two given vectors contain altogether six components which enable the calculation of the six Keplerian parameters.

Position vectors are preferred for orbit determination since they are more accurate than velocity vectors. Let us assume that two position vectors $\underline{\varrho}^S(t_1)$ and $\underline{\varrho}^S(t_2)$ at epochs t_1 and t_2 are available. These data correspond to boundary values in the solution of the basic differential equation of second order, cf. Eq. (4.1). An approximate method for the derivation of the Keplerian parameters makes use of initial values defined for an averaged epoch $t = \frac{1}{2}(t_1 + t_2)$:

$$
\begin{aligned}
\underline{\varrho}^S(t) &= \frac{\underline{\varrho}^S(t_2) + \underline{\varrho}^S(t_1)}{2} \\
\underline{\dot{\varrho}}^S(t) &= \frac{\underline{\varrho}^S(t_2) - \underline{\varrho}^S(t_1)}{t_2 - t_1} .
\end{aligned}
\tag{4.49}
$$

The rigorous solution starts with the computation of the geocentric distances

$$
\begin{aligned}
r_1 &= r(t_1) = \|\underline{\varrho}^S(t_1)\| \\
r_2 &= r(t_2) = \|\underline{\varrho}^S(t_2)\| .
\end{aligned}
\tag{4.50}
$$

The unit vector $\underline{e}_3$, orthogonal to the orbital plane, is obtained from a vector product by

$$\underline{e}_3 = \frac{\varrho^S(t_1) \times \varrho^S(t_2)}{\|\varrho^S(t_1) \times \varrho^S(t_2)\|} \tag{4.51}$$

and produces the longitude ℓ and the inclination angle i, cf. Eqs. (4.11) and (4.13). As demonstrated earlier, the argument of the latitude $u = \omega + v$ is defined as the angle between the satellite's position and the ascending node vector $\underline{k} = (\cos \ell, \sin \ell, 0)^T$. Hence, the relation

$$r_i \cos u_i = \underline{k} \cdot \varrho^S(t_i), \qquad i = 1, 2 \tag{4.52}$$

holds from which the u_i with $u_2 > u_1$ can be deduced uniquely. Now, there are two equations, cf. Eq. (4.7),

$$r_i = \frac{a(1 - e^2)}{1 + e \cos(u_i - \omega)}, \qquad i = 1, 2 \tag{4.53}$$

where the parameters a, e, ω are unknown. The system can be solved for a and e after assigning a preliminary value such as the nominal one to ω, the argument of the perigee. With the assumed ω and the u_i the true anomalies v_i and subsequently the mean anomalies M_i are obtained. Therefore, the mean angular velocity n can be calculated twice by the formulas

$$n = \sqrt{\frac{\mu}{a^3}} = \frac{M_2 - M_1}{t_2 - t_1}, \tag{4.54}$$

cf. Eqs. (4.2) and (4.3). The equivalence is achieved by varying ω. This iterative procedure is typical for boundary value problems. Finally, the epoch of perigee passage T_0 follows from the relation

$$T_0 = t_i - \frac{M_i}{n}. \tag{4.55}$$

For a numerical solution of the boundary value problem assume two position vectors $\varrho(t_1)$ and $\varrho(t_2)$, both represented in the earth-fixed equatorial system $\underline{X}_i$ and with $\Delta t = t_2 - t_1 = 1$ hour:

$$\varrho(t_1) = (11\,465, \quad 3\,818, \quad -20\,923)^T \quad [\text{km}]$$

$$\varrho(t_2) = (\quad 5\,220, \quad 16\,754, \quad -18\,421)^T \quad [\text{km}].$$

The application of the Eqs. (4.50) through (4.55), apart from rounding errors, results in the following set of parameters for the associated Kepler ellipse: $a = 26\,000$ km, $e = 0.1$, $\omega = -140°$, $i = 60°$, $\ell = 110°$, and $T_0 = t_1 - 1^h.3183$.

If there are redundant range observations, the parameters of an instantaneous Kepler ellipse can be improved. The position vector ϱ_0^S associated to the reference ellipse can be computed, and each observed range gives rise to an equation

$$\varrho = \varrho_0 + d\varrho = \|\underline{\varrho}_0^{\cdot S} - \underline{\varrho}_R\| + \frac{\underline{\varrho}_0^S - \underline{\varrho}_R}{\|\underline{\varrho}_0^S - \underline{\varrho}_R\|} \cdot d\underline{\varrho}^S . \tag{4.56}$$

The vector $d\underline{\varrho}^S$ can be expressed as a function of the Keplerian parameters, cf. Eq. (4.29). Thus, Eq. (4.56) actually contains the differential increments for the six orbital parameters.

An orbit improvement is often performed in the course of GPS data processing when in addition to terrestrial position vectors ϱ_R the solution also allows for the determination of the increments dp_{0i}. Of course the procedure becomes unstable or even fails for small networks. Sometimes only three degrees of freedom are assigned to the orbit by a shift vector. This procedure is called orbit relaxation.

4.3.2 Perturbed orbit

Analytical solution. As known from previous sections, the perturbed motion is characterized by temporal variations of the orbital parameters. The analytical expressions for these variations are given by the Eqs. (4.37) or (4.39).

In order to be applicable for Lagrange's equations, the disturbing potential must be expressed as a function of the Keplerian parameters. For the effects of the nonsphericity of the earth the Legendre functions in Eq. (4.40) must be transformed which first has been performed by Kaula (1966). The resulting relation is

$$R = \sum_{n=2}^{\infty} \sum_{m=0}^{n} A_n(a) \sum_{p=0}^{n} F_{nmp}(i)$$
$$\cdot \sum_{q=-\infty}^{\infty} G_{npq}(e) \, S_{nmpq}(\omega, \Omega, M; \Theta_0, J_{nm}, K_{nm}) . \tag{4.57}$$

Recall that n denotes the degree and m the order of the spherical harmonics in the disturbing potential development. Each of the functions A_n, F_{nmp}, G_{npq} contains only one parameter of a Kepler ellipse. However, the function S_{nmpq} is composed of several parameters and can be expressed by

$$\begin{aligned} S_{nmpq} &= J_{nm} \cos \psi + K_{nm} \sin \psi \qquad \text{for } n-m \text{ even} \\ S_{nmpq} &= -K_{nm} \cos \psi + J_{nm} \sin \psi \qquad \text{for } n-m \text{ odd}. \end{aligned} \tag{4.58}$$

Table 4.4. Perturbations due to the earth's gravity field

Parameter	Secular	Long-period	Short-period
a	no	no	yes
e	no	yes	yes
i	no	yes	yes
Ω	yes	yes	yes
ω	yes	yes	yes
M	yes	yes	yes

The relation of the frequency $\dot{\psi}$ associated to the argument ψ is

$$\dot{\psi} = (n - 2p)\,\dot{\omega} + (n - 2p + q)\,\dot{M} + m\,(\dot{\Omega} - \dot{\Theta}_0) \qquad (4.59)$$

which is a measure for the spectrum of the perturbations.

The conditions $(n - 2p) = (n - 2p + q) = m = 0$ lead to $\dot{\psi} = 0$ and thus to secular variations. Because of $m = 0$, they are caused by zonal harmonics. If $(n - 2p) \neq 0$, then the variations depend on $\dot{\omega}$ and are, therefore, generally long-periodic. Finally, the conditions $(n - 2p + q) \neq 0$ and/or $m \neq 0$ result in short-period variations. The integer value $(n - 2p + q)$ gives the frequency in cycles per revolution and m the frequency in cycles per day.

A rough overview of the frequency spectrum in the Keplerian parameters due to the gravity field of the earth is given in Table 4.4. Summarizing, one can state that the even degree zonal coefficients produce primarily secular variations and the odd degree zonal coefficients give rise to long-periodic perturbations. The tesseral coefficients are responsible for short-periodic terms. From Table 4.4 one can read that short-period variations occur in each parameter. With the exception of the semimajor axis, the parameters are also affected by long-periodic perturbations. However, secular effects are only contained in Ω, ω, M. The analytical expression for the secular variations in these parameters due to the oblateness term J_2 is given as an example:

$$\dot{\Omega} = -\frac{3}{2}\,n\,a_E^2\,\frac{\cos i}{a^2\,(1 - e^2)^2}\,J_2$$

$$\dot{\omega} = \frac{3}{4}\,n\,a_E^2\,\frac{5\cos^2 i - 1}{a^2\,(1 - e^2)^2}\,J_2 \qquad (4.60)$$

$$\dot{m} = \frac{3}{4}\,n\,a_E^2\,\frac{3\cos^2 i - 1}{a^2\,\sqrt{(1 - e^2)^3}}\,J_2\,.$$

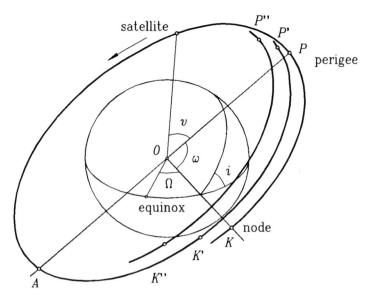

Fig. 4.3. Secular perturbations caused by the oblateness term J_2

The first equation describes the regression of the node in the equatorial plane, the second equation expresses the rotation of the perigee, and the third equation contributes to the variation of the mean anomaly by $\dot{M} = n + \dot{m}$. In the case of GPS satellites, the numerical values $\dot{\Omega} \approx -0.^{\circ}03$ per day, $\dot{\omega} \approx 0.^{\circ}01$ per day, and $\dot{m} \approx 0$ are obtained. The result for $\dot{m}$ is verified immediately since for the nominal inclination $i = 55°$ the term $3\cos^2 i - 1$ becomes approximately zero. A graphical representation of the secular perturbations is given in Fig. 4.3. Special attention must be paid to resonance effects which occur when the period of revolution corresponds to a harmonic in the gravity potential. This is why the GPS satellites are raised into slightly higher orbits to preclude an orbital period very close to half a sidereal day.

The tidal potential also has a harmonic representation. Analogously, the tidal perturbations can be analytically modeled. This was done first by Kozai (1959) and, analogously to the earth's potential effect, has led to analytical expressions for the secular variations of the node's right ascension Ω and of the perigee's argument ω. The reader is referred to the publication of Kozai (1959) for formulas specified.

Numerical solution. If the disturbing acceleration cannot be expressed in analytical form, one has to apply numerical methods for the solution. Therefore, in principle, with initial values such as the position and velocity vectors

$\underline{\varrho}(t_0)$ and $\underline{\dot{\varrho}}(t_0)$ at a reference epoch t_0, a numerical integration of Eq. (4.30) could be performed. This simple concept can be improved by the introduction of a Kepler ellipse as a reference. By this means only the smaller difference between the total and the central acceleration must be integrated. The integration results in an increment $\delta\varrho$ which, when added to the position vector computed for the reference ellipse, gives the actual position vector.

The second-order differential equation is usually transformed to a system of two differential equations of the first order for the numerical integration. This system is given by

$$\underline{\dot{\varrho}}(t) = \underline{\dot{\varrho}}(t_0) + \int_{t_0}^{t} \underline{\ddot{\varrho}}(t_0)\,dt$$

$$= \underline{\dot{\varrho}}(t_0) + \int_{t_0}^{t} \left[d\underline{\ddot{\varrho}}(t_0) - \frac{\mu}{\varrho^3(t_0)}\,\underline{\varrho}(t_0) \right] dt \qquad (4.61)$$

$$\underline{\varrho}(t) = \underline{\varrho}(t_0) + \int_{t_0}^{t} \underline{\dot{\varrho}}(t_0)\,dt \ .$$

The numerical integration of this coupled system can be performed by applying the standard Runge-Kutta algorithm, cf. for example Kreyszig (1968), p. 89. This method is shortly outlined. Let $y(x)$ be a function defined in the interval $x_1 \leq x \leq x_2$ and denote by $y' = dy/dx$ its first derivative with respect to the argument x. The general solution of the ordinary differential equation of the first order

$$y' = \frac{dy}{dx} = y'(y, x) \qquad (4.62)$$

follows from integration and the particular solution is found after assigning the given numerical initial value $y_1 = y(x_1)$ to the integration constant. For the application of numerical integration, first the integration interval is subdivided into n equal and sufficiently small $\Delta x = (x_2 - x_1)/n$ where n is an arbitrary integer > 0. Then, the difference between successive functional values is obtained by the weighted mean

$$\Delta y = y(x + \Delta x) - y(x)$$

$$= \frac{1}{6}\left[\Delta y^{(1)} + 2\left(\Delta y^{(2)} + \Delta y^{(3)}\right) + \Delta y^{(4)} \right] \qquad (4.63)$$

where

$$\Delta y^{(1)} = y'(y,\ x)\,\Delta x$$

$$\Delta y^{(2)} = y'(y + \frac{\Delta y^{(1)}}{2},\ \ x + \frac{\Delta x}{2})\,\Delta x$$

$$\Delta y^{(3)} = y'(y + \frac{\Delta y^{(2)}}{2},\ \ x + \frac{\Delta x}{2})\,\Delta x$$

$$\Delta y^{(4)} = y'(y + \Delta y^{(3)},\ \ x + \Delta x\,)\,\Delta x\,.$$

Hence, starting with the initial value y_1 for the argument x_1, the function can be calculated for the successive argument $x_1 + \Delta x$ and so on.

Numerical methods can also be applied to the integration of the disturbing equations of Lagrange or Gauss. These equations have the advantage of being differentials of the first order and, therefore, must only be integrated once over time.

Let us conclude this paragraph with an example for numerical integration. Assume the ordinary first-order differential equation $y' = y - x + 1$ with the initial value $y_1 = 1$ for $x_1 = 0$. The differential equation should be solved for the argument $x_2 = 1$ with increments $\Delta x = 0.5$. Starting with the initial values for the first interval, successively the values $\Delta y^{(1)} = 1.000$, $\Delta y^{(2)} = 1.125$, $\Delta y^{(3)} = 1.156$, and $\Delta y^{(4)} = 1.328$ are obtained. The corresponding weighted mean thus is $\Delta y = 1.148$. Replacing in the second interval the initial values by $x = x_1 + \Delta x = 0.5$ and $y = y_1 + \Delta y = 2.148$ and proceeding in an analogous manner as before yields $\Delta y = 1.569$ and thus the final result $y_2 = y(1) = y + \Delta y = 2.717$. Note that the function $y = e^x + x$, satisfying the differential equation, gives a true value $y(1) = 2.718$. With an increment $\Delta x = 0.1$ the numerical integration would provide an accuracy in the order of 10^{-6}.

4.4   Orbit dissemination

4.4.1   Tracking networks

Objectives and strategies. At present the broadcast ephemerides are accurate to about 5 m for Block I satellites with cesium clocks under the assumption of three uploads per day as it was demonstrated by Remondi and Hofmann-Wellenhof (1989a). For the Block II satellites, the accuracy for nonauthorized users can be degraded to an unspecified amount believed to be in the 30 to 50 m range by the implementation of SA. However, in the near future an orbital accuracy of about 20 cm is required for specific missions such as TOPEX/Poseidon or for such investigations which require an accuracy at the level of 10^{-9}. Moreover, the geodetic community will become independent from the policy of the Department of Defense (DoD) with a global GPS tracking network. Therefore, the need for civil tracking networks for orbit determination is evident.

Global networks result in higher accuracy and reliability of the orbits compared to those determined from regional networks. The tie of the orbital system to terrestrial reference frames is achieved by the collocation of GPS receivers with VLBI and SLR sites. The distribution of the GPS sites is essential to achieve the highest accuracy. Two different approaches can be compared. In the first case, the sites are regularly distributed around the globe; in the second case each network site is surrounded by a cluster of additional points in order to facilitate ambiguity resolution and thus to strengthen the solution for the orbital parameters by a factor three to five, cf. Counselman and Abbot (1989). If a configuration is aspired where at least two satellites can be tracked simultaneously any time from two sites, the minimum number of sites in a global network is six, cf. Lichten and Neilan (1990). A densified version should contain approximately 20 sites with clusters around them.

Examples. Apart from the GPS control segment, several networks have been established for orbit determination. Without mentioning the numerous networks on a regional or even continental scale such as the Australian GPS orbit determination pilot project reported by Rizos et al. (1987), some examples of global networks are given.

The Cooperative International GPS Network (CIGNET) is being operated by the U.S. National Geodetic Survey (NGS) with tracking stations located at VLBI sites. In early 1991, about 20 stations, cf. Fig. 4.4 were participating. These tracking sites are listed in Table 4.5, cf. Chin (1991). Twelve of the CIGNET participants (marked by an asterisk in Table 4.5) download their tracking data every night to the processing center located at the NGS facility in Rockville, Maryland. The NGS processes these data and makes it available on the following morning to participants by means of a file server computer connected to modems. Data from the remaining stations are received at NGS with a delay of two to seven days and are processed and made available within one day.

The CIGNET is different from the official tracking network in several aspects. First and most important, these stations observe and record the phase as well as the code range data from all satellites. They thus observe the C/A-code and P-code pseudoranges on the $L1$ frequency and the P-code pseudorange on the $L2$ frequency as well as the carrier phase on both frequencies. When A-S is invoked, these sites will be unable to track the

Table 4.5. CIGNET sites (status in early 1991)

	Site	Country
	Hartebeesthoek	South Africa
*	Hobart	Australia
*	Kokee Park	Hawaii, USA
	Madrid	Spain
*	Mojave/Goldston	California, USA
	Natal	Brazil
*	Onsala	Sweden
	Port Harcourt	Nigeria
*	Richmond	Florida, USA
	Santiago	Chile
	Tidbinbilla	Australia
*	Townsville	Australia
*	Tromsø	Norway
*	Tsukuba-Kashima	Japan
	Useda	Japan
*	Wellington	New Zealand
*	Westford	Massachusetts, USA
*	Wettzell	Germany
	Yaragadee	Australia
*	Yellowknife	Canada

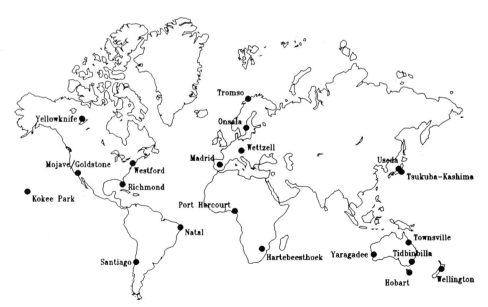

Fig. 4.4. CIGNET sites (status in early 1991)

P-code using their present equipment. A second major difference in the CIGNET is the fact that the tracking data are immediately available to users, whereas the official precise ephemerides is subject to a two-week delay.

Participants (and possibly others) can access the tracking data from CIGNET and, if they are equipped with appropriate programs, they would be able to compute their own ephemerides. Another application of the tracking data is to combine it with data from a local survey and use orbit relaxation software to compute the coordinates of local points. The NGS has tested this concept using data from the U.S. CIGNET stations and obtained a few centimeters accuracy for points in the mid-continent.

The GPS Bulletin prepared and disseminated bimonthly by the NGS provides tracking statistics, field notes, receiver upgrades, improvement of the network, and other information concerning the CIGNET.

The first Global Orbit Tracking Experiment (GOTEX) was performed in the fall 1988 and aimed at the collocation of GPS at existing VLBI and SLR sites. The data should permit a close connection between the WGS-84 and the VLBI/SLR systems. About 25 sites, distributed worldwide, were occupied during the three weeks of the campaign. In the meantime the data were reformatted into the CIGNET format and may be obtained from the NGS.

In 1990, the International Association of Geodesy (IAG) decided to install an International GPS Geodynamics Service (IGS). After a test campaign in 1992, routine activities should begin in 1993. The main purpose of this service will be dedicated to geodynamical applications which require highest accuracy. Another task of the service is the establishment of an orbit determination activity similar to CIGNET. The network is planned to consist of a permanent core network with about 25 control stations supplemented by more than 100 intermittent fiducial stations. Further details can be found in Mueller (1991).

4.4.2 Ephemerides

Almanac data. The purpose of the almanac data is to provide the user with less precise data to facilitate receiver satellite search or for planning tasks such as the computation of visibility charts. The almanac data is transmitted via the satellite message, cf. Sect. 5.1.2, and essentially contains parameters for the orbit representation, satellite clock correction parameters, and some other information, cf. Table 4.6. The parameter ℓ_0 denotes the difference between the node's right ascension at epoch t_a and the Greenwich sidereal time at t_0, the beginning of the current GPS week. The reduction of the

Table 4.6. Almanac data

Parameter	Explanation
ID	Satellite PRN number
HEALTH	Satellite health status
WEEK	Current GPS week
t_a	Reference epoch in seconds within the current week
$\sqrt{a}$	Square root of semimajor axis
e	Eccentricity
M_0	Mean anomaly at reference epoch
ω	Argument of perigee
δi	Offset from nominal inclination of 55°
ℓ_0	Longitude of the node at weekly epoch
$\dot{\Omega}$	Rate of the node's right ascension
a_0	Clock phase bias
a_1	Clock frequency bias

Keplerian parameters to the observation epoch t is obtained by the formulas

$$M = M_0 + n\,(t - t_a)$$

$$i\ = 55° + \delta i \tag{4.64}$$

$$\ell\ = \ell_0 + \dot{\Omega}\,(t - t_a) - \omega_E\,(t - t_0)$$

with ω_E being the earth's angular velocity. The other three Keplerian parameters a, e, ω remain unchanged. Note that in the formula for ℓ in Eq. (4.64) the second right-hand side term considers the node's regression and the third term expresses the uniform change in the sidereal time since epoch t_0. An estimate for the satellite clock bias is given by

$$\delta^S = a_0 + a_1\,(t - t_a)\,. \tag{4.65}$$

Broadcast ephemerides. The broadcast ephemerides are based on observations at the five monitor stations of the GPS control segment. The most recent of these data are used to compute a reference orbit for the satellites. Additional tracking data are entered into a Kalman filter and the improved orbits are used for extrapolation, cf. Wells et al. (1987). As reported by Remondi and Hofmann-Wellenhof (1989b), these orbital data are accurate to approximately 5 m based on three uploads per day; with a single daily update one might expect an accuracy of 10 m. The Master Control Station

Table 4.7. Broadcast ephemerides

Parameter	Explanation
AODE	Age of ephemerides data
t_e	Ephemerides reference epoch
$\sqrt{a},\, e,\, M_0,$ $\omega_0,\, i_0,\, \ell_0$	Keplerian parameters at t_e
Δn	Mean motion difference
$\dot{i}$	Rate of inclination angle
$\dot{\Omega}$	Rate of node's right ascension
$C_{uc},\, C_{us}$	Correction coefficients (argument of perigee)
$C_{rc},\, C_{rs}$	Correction coefficients (geocentric distance)
$C_{ic},\, C_{is}$	Correction coefficients (inclination)

is responsible for the computation of the ephemerides and the upload to the satellites. Essentially, the ephemerides contain six parameters to describe a smoothed Kepler ellipse at a reference epoch and some secular and periodic correction terms. The most recent parameters listed in Table 4.7 are broadcast every hour and should only be used during the prescribed period of approximately four hours to which they refer.

The perturbation effects due to the nonsphericity of the earth, the direct tidal effect, and the solar radiation pressure are considered by the last nine terms in Table 4.7. In order to compute the satellite's position at the observation epoch, apart from the parameters a and e, the following quantities are needed

$$M = M_0 + \left[\sqrt{\frac{\mu}{a^3}} + \Delta n\right] (t - t_e)$$

$$\ell = \ell_0 + \dot{\Omega}\,(t - t_e) - \omega_E\,(t - t_0)$$

$$\omega = \omega_0 + C_{uc} \cos(2u) + C_{us} \sin(2u) \qquad (4.66)$$

$$r = r_0 + C_{rc} \cos(2u) + C_{rs} \sin(2u)$$

$$i = i_0 + C_{ic} \cos(2u) + C_{is} \sin(2u) + \dot{i}\,(t - t_e)$$

with $u = \omega + v$ being the argument of latitude. The geocentric distance r_0 is calculated by Eq. (4.7) using a, e, E at the observation epoch. With the reference epoch t_e the computation of ℓ is analogous to that in Eq. (4.64).

Precise ephemerides. The precise ephemerides are based on observed data obtained in tracking networks and are computed by several institutions. The results are available some weeks after data collection and are expressed as satellite positions and velocities at equidistant epochs. As an example, the NGS format according to Remondi (1989) is described here.

The complete NGS format consists of a header containing general informations (epoch interval, orbit type, etc.) followed by the data section for successive epochs. These data are reported for each satellite and represent the position vector ϱ [km] and the velocity vector $\dot{\varrho}$ [km·s^{-1}].

The reduced NGS format contains only positions since it is shown by Remondi (1989) and Remondi (1991b) that, in general, velocity is superfluous. The velocity information can be computed from the position data, with sufficient precision, by differentiating an interpolating polynomial. Hence, only the half storage amount is needed compared to the complete format.

Until recently the NGS results have been based on dual frequency P-code range data obtained from the CIGNET. More recently the NGS has switched to a carrier phase observable. The orbits are computed using the IERS station coordinates which approximate the WGS-84 values. The internal precision of the orbits is at the level of a few parts in 10^{-7}. NGS disseminates the data as ASCII and binary files. The latter is formatted for IBM compatible PC's, saves storage capacity, and facilitates electronic data transfer. The position and velocity vectors between the given epochs are obtained by interpolation where the Lagrange interpolation based on polynomial base functions is used.

Note that Lagrange interpolation is also applicable to variable epoch series and that the coefficients determined can be applied to considerably longer series without updating the coefficients. This interpolation method is a fast procedure and can be programmed easily. Extensive studies by Remondi (1989) concluded that for GPS satellites a 30-minute epoch interval and a 9th order interpolator suffices for an accuracy of about 10^{-8}. A more recent study by Remondi (1991b) using a 17th order interpolator demonstrates that millimeter-level (10^{-10}) can be achieved based on a 40-minute epoch interval.

For those not familiar with Lagrange interpolation, the principle of this method and a numerical example is given, cf. for example Moritz (1977). Let the epochs t_j, $j = 0, \ldots, n$ be where the associated functional values $f(t_j)$ are given. Then,

$$
l_j(t) = \frac{(t - t_0)(t - t_1) \cdots (t - t_{j-1})(t - t_{j+1}) \cdots (t - t_n)}{(t_j - t_0)(t_j - t_1) \cdots (t_j - t_{j-1})(t_j - t_{j+1}) \cdots (t_j - t_n)}
$$

$$(4.67)$$

is the definition of the corresponding base functions $l_j(t)$ of degree n related to an arbitrary epoch t. The interpolated functional value at epoch t follows from the summation

$$f(t) = \sum_{j=0}^{n} f(t_j) l_j(t). \tag{4.68}$$

The following numerical example assumes the functional values $f(t_j)$ given at the epochs t_j:

$$\underline{f} = (f(t_j)) = (13, 17, 85)^T$$

$$\underline{t} = (t_j) \quad = (-3, 1, 5)^T.$$

The base functions are polynomials of second degree

$$l_0(t) = \frac{(t - t_1)(t - t_2)}{(t_0 - t_1)(t_0 - t_2)} = \frac{1}{32}(t^2 - 6t + 5)$$

$$l_1(t) = \frac{(t - t_0)(t - t_2)}{(t_1 - t_0)(t_1 - t_2)} = -\frac{1}{16}(t^2 - 2t - 15)$$

$$l_2(t) = \frac{(t - t_0)(t - t_1)}{(t_2 - t_0)(t_2 - t_1)} = \frac{1}{32}(t^2 + 2t - 3)$$

and according to Eq. (4.68) the interpolated value for $t = 4$ is $f(t) = 62$. The result is immediately verified since the given functional values were generated by the polynomial $f(t) = 2t^2 + 5t + 10$.

5. Satellite signal

5.1 Signal structure

5.1.1 Physical fundamentals

Operational satellite geodesy is based upon data transmitted from the satellite to the user by means of electromagnetic waves. Some of the factors describing the physical behaviour of these waves are presented in Table 5.1 together with their symbols and dimensions. Note that integer cycles are equivalent to multiples of 2π radians. Another unit for cycles per second (cps) is Hertz (Hz) or decimal multiples according to the rules of the International System of Units (ISU). Equation (5.1) relates the parameters shown in Table 5.1:

$$f = 2\pi \frac{1}{P} = \frac{c}{\lambda} .\tag{5.1}$$

The instantaneous circular frequency f is also defined by the derivation of the phase φ with respect to time

$$f = \frac{d\varphi}{dt} \tag{5.2}$$

and the phase is obtained from integrating the frequency between the epochs t_0 and t

$$\varphi = \int_{t_0}^{t} f \, dt . \tag{5.3}$$

Assuming a constant frequency, setting the initial phase $\varphi(t_0) = 0$, and taking into account the time span t_ϱ which the signal needs to propagate

Table 5.1. Physical quantities

Quantity	Symbol	Dimension
Circular frequency	f	$\text{cycle} \cdot \text{s}^{-1}$
Phase	φ	cycle
Wavelength	λ	$\text{m} \cdot \text{cycle}^{-1}$
Period	P	s
Speed of light	c	$\text{m} \cdot \text{s}^{-1}$

through the distance ϱ from the emitter to the receiver yields the phase
equation for electromagnetic waves as observed at the receiving site:

$$\varphi = f\,(t - t_\varrho) = f\,(t - \frac{\varrho}{c}). \tag{5.4}$$

For a numerical example consider an electromagnetic wave with frequency
$f = 1.5\,\text{GHz}$ and calculate the instantaneous phase at an antenna $20\,000\,\text{km}$
distant from the transmitter source. With $c = 3 \cdot 10^5\,\text{km·s}^{-1}$ a (continuous)
phase of exactly 10^8 cycles results. The observable (fractional) phase within
one cycle thus is zero.

In the case of a moving emitter or a moving receiver, the received frequency is Doppler shifted. This means that the received frequency f_r differs
from the emitted frequency f_e by an amount Δf which, apart from relativistic effects, is proportional to the radial velocity $v_\varrho = d\varrho/dt = \dot{\varrho}$ of the
emitter with respect to the receiver:

$$\Delta f = (f_r - f_e) = -\frac{1}{c}\,v_\varrho\,f_e. \tag{5.5}$$

It can be verified by Eq. (4.9) that GPS satellites are orbiting with the
mean velocity $\dot{r} = n\,a \approx 3.9\,\text{km·s}^{-1}$. With respect to a stationary terrestrial receiver there is a zero radial velocity and thus no Doppler effect at the
epoch of closest approach. The maximum radial velocity appears when the
satellite crosses the horizon and is $0.9\,\text{km·s}^{-1}$. Assuming a transmitted frequency $f_e = 1.5\,\text{GHz}$, the Doppler frequency shift is $\Delta f = 4.5 \cdot 10^3\,\text{Hz}$. This
frequency shift leads to a phase change of 4.5 cycles after one millisecond
which corresponds to a change in range of 90 cm. The result can be verified
by multiplying the radial velocity or range rate with the time span of one
millisecond.

5.1.2 Components of the signal

General remarks. The oscillators on board the satellites generate a fundamental frequency f_0 with a stability in the range of 10^{-13} over one day for
the Block II satellites. Two carrier signals in the L-band, denoted $L1$ and
$L2$, are generated by integer multiplications of f_0. These carriers are modulated by codes to provide satellite clock readings to the receiver and to
transmit information such as the orbital parameters. The codes consist of
a sequence with the states $+1$ or -1, corresponding to the binary values 0
or 1. The so-called biphase modulation is performed by a $180°$ shift in the
carrier phase whenever a change in the state occurs, cf. Fig. 5.1.

The components of the signal and their frequencies are summarized in
Table 5.2. Note that the nominal fundamental frequency f_0 is intentionally

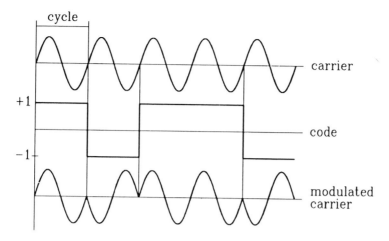

Fig. 5.1. Biphase modulation of carrier wave

Table 5.2. Components of the satellite signal

Component	Frequency (MHz)		
Fundamental frequency	f_0	$= 10.23$	
Carrier $L1$	$154 f_0$	$= 1\,575.42$	($\hat{=}$ 19.0 cm)
Carrier $L2$	$120 f_0$	$= 1\,227.60$	($\hat{=}$ 24.4 cm)
P-code	f_0	$= 10.23$	
C/A-code	$f_0/10$	$= 1.023$	
Navigation message	$f_0/204\,600$	$= 50 \cdot 10^{-6}$	

reduced by about 0.005 Hz to compensate for relativistic effects. More details on this subject are provided in Sect. 6.4.3.

Two codes are used for the satellite clock readings, both characterized by a pseudorandom noise (PRN) sequence. The coarse/acquisition code (C/A-code) has the frequency $f_0/10$ and is repeated every millisecond. The precision code (P-code) has the frequency f_0 and is repeated approximately once every 266.4 days. The coding of the navigation message requires 1\,500 bits and, at the frequency of 50 Hz, is transmitted in 30 seconds.

Both the $L1$ and $L2$ carrier are modulated by the P-code. The C/A-code is placed on the $L1$ carrier in phase quadrature (i.e., 90° offset) with the P-code. Denoting unmodulated carrier by $Li(t) = a_i \cos(f_i t)$ and the state sequences of the P-code, the C/A-code, and the navigation message by $P(t)$, $C/A(t)$, and $D(t)$, respectively, the modulated carriers are represented by the equations, cf. Spilker (1980),

$$L1(t) = a_1 \, P(t) \, D(t) \cos(f_1 \, t) + a_1 \, C/A(t) \, D(t) \sin(f_1 \, t)$$

$$L2(t) = a_2 \, P(t) \, D(t) \cos(f_2 \, t) \,.$$

(5.6)

Note that the broadcast signal is a spread spectrum signal which ensures more communication security.

Pseudorandom noise codes. The generation of the PRN sequences in the P-code and the C/A-code is based on the use of hardware devices called tapped feedback shift registers, cf. Wells et al. (1987), Sect. 6.4. As illustrated in Fig. 5.2, such a device consists of a number of storage cells (here labeled 1–5) comprising one bit. Associated with each clock pulse is a shift of the bits to the right where the content of the rightmost cell is read as output. The new value of the leftmost cell is determined by the binary sum of two defined cells where a binary 0 is set if the bits in these two cells are equal. The choice of the defining cells is arbitrary and determines the property of the resulting code.

In order to make the procedure more clear consider cells 2 and 3 as defining pair in Fig. 5.2. After one clock pulse the initial state would deliver a binary 0 as output and a binary 1 would be set in cell 1 of the successive state. Continuing in an anologous way, the bit sequence 1101110010... would be delivered.

The bits of the PRN sequences are often called chips to underscore that these codes do not carry data.

The C/A-code is generated by the combination of two 10-bit tapped feedback shift registers where the output of both registers are added again by binary operation to produce the code sequence. The codes of the two registers are not classified and the C/A-code is available to civilian users. A unique code is assigned to each satellite depending on the defining cells. The frequency of 1.023 MHz and the repetition rate of 1 millisecond results in a code length of 1 023 chips. Hence, the time interval between two chips is just under 1 microsecond which approximately corresponds to a 300 m chip length.

Number of cell	1	2	3	4	5
Initial state	1	0	1	1	0
Successive state	1	1	0	1	1

Fig. 5.2. Principle of a tapped feedback shift register

5 * $S1$	0 1 0 0 1 0 0 1 0 0 1 0 0 1 0
3 * $S2$	1 0 1 1 0 1 0 1 1 0 1 0 1 1 0
Combination	1 1 1 1 1 1 0 0 1 0 0 0 1 0 0

Fig. 5.3. Principle of P-code generation

The P-code is also not classified and originates from a combination of two bit sequences, again each generated by two registers. The first bit sequence repeats every 1.5 second and, because of the frequency f_0, has a length of $1.5345 \cdot 10^7$ bits. The second sequence has 37 more bits. The combination of both sequences thus results in a code with approximately $2.3547 \cdot 10^{14}$ bits which corresponds to a time span of approximately 266.4 days as stated earlier. The total code length is partitioned into 37 unique one-week segments and each segment is assigned to a satellite defining its PRN number. The codes are restarted at the beginning of every GPS week at Saturday midnight. The chip length of the P-code is about 30 m. The reading of the satellite clock at any time is obtained by adding the number of chips within the actual 1.5-second interval to the so-called Z-count which is transmitted in the navigation message and represents a multiple number of 1.5-second intervals since the beginning of the current GPS week. In order to protect the P-code against spoofing (i.e., the deliberate transmission of incorrect information by an adversary), the code can be encrypted by invoking antispoofing, cf. Sect. 2.2.3. This procedure converts the P-code to the Y-code which is only useable when the secret conversion algorithm is accessible. In order to illustrate the P-code generation, an example is given, cf. Fig. 5.3. Consider the two sequences $S1$ and $S2$ containing the bit sequences 010 and 10110 respectively. The combination of 5 * $S1$ with 3 * $S2$ using binary addition results in a code with a length of 15 chips. Since $S2$ comprises 2 bits more than $S1$, the combined code can be partitioned into two unique segments where their length is two times the length of $S1$.

Navigation message. The navigation message essentially contains information on the satellite clock, the satellite orbit, the satellite health status, and various correction data. As shown schematically in Table 5.3, the total message consisting of 1 500 bits is subdivided into five subframes. One subframe is transmitted in 6 seconds and contains 10 words with 30 bits. The transmission time needed for a word is therefore 0.6 seconds.

Each subframe starts with the telemetry word (TLM) containing a synchronization pattern and some diagnostic messages. The second word in each

Table 5.3. Scheme of navigation message

	Number of bits	Transmission time
Total message	1 500	30 seconds
Subframe (1–5)	300	6 seconds
Word (1–10)	30	0.6 seconds

subframe is the hand-over word (HOW). Apart from an identification, this word contains a number which, when multiplied by four, gives the Z-count for the next subframe.

The first subframe contains the coefficients for a quadratic polynomial to model the satellite clock correction, indicators on the age of the data, and various flags.

The second and third subframe transmit the broadcast ephemerides of the satellite as described in Sect. 4.4.2.

The contents of the fourth and the fifth subframe are changed in every message and have a repetition rate of 25. The total information thus is packed into 25 pages and needs 12.5 minutes for transmission. Many pages of the fourth subframe are reserved for military use; the rest contains information on the ionosphere, UTC data, various flags, and the almanac data (i.e., low-accuracy orbital data) for all the satellites exceeding the nominal constellation. The pages of the fifth subframe are mainly dedicated to the almanac data and the health status for the first 24 satellites in orbit. The pages of the fourth and fifth subframe are broadcast by each satellite. Hence, by tracking only one satellite one gets information on the almanac data of all the other satellites in orbit.

For more information concerning the data of the navigation message the reader is referred to Sect. 4.4 on orbit dissemination. A detailed description of the data format is given in van Dierendonck et al. (1980), Rockwell International Corporation (1984), and Nieuwejaar (1988).

5.2   Signal processing

5.2.1   General remarks

The signal emitted from the satellite is represented by Eq. (5.6) and contains three components in the symbolic form $(L1, C/A, D)$, $(L1, P, D)$, and $(L2, P, D)$. The processing of the signal at the GPS receiver aims at the

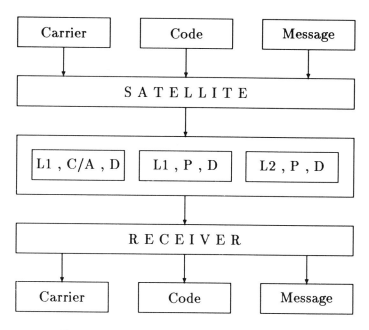

Fig. 5.4. Principle of signal processing

recovery of the signal components, including the reconstruction of the carrier
wave and the extraction of the codes for the satellite clock readings and the
navigation message. The principle is illustrated in Fig. 5.4.

5.2.2 Receiver

General features. Presently, there are more than 100 receivers on the market used for different purposes (navigation, surveying, time transfer) and
with different features. Despite of this variety, all the receivers show certain
common principles which are presented in this section.

Following Langley (1991), the receiver unit contains elements for signal
reception and signal processing. The basic conception is shown in Fig. 5.5.

An omnidirectional antenna receives the signals of all satellites above the
horizon and, after preamplification, transmits them to the radio frequency
section. It is worth noting that the signals are very resistant to interferences
since the PRN codes are unique for each satellite and have very low cross
correlations. The antenna may be designed for only the primary carrier $L1$
or for both $L1$ and $L2$ carriers. Today, most of the antennas on the market
are microstrip antennas. The important design criterion of the antenna is
the sensitivity of the phase center. The electronic center should be close to

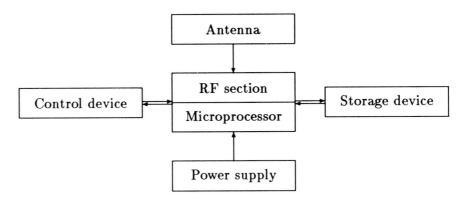

Fig. 5.5. Basic conception of a receiver unit

its physical center and should be insensitive with respect to rotation and inclination. This becomes particularly important in kinematic applications where the antenna is moved while in operation. Additionally, the antenna should have a gain pattern which filters low elevation or multipath signals. Additional details on these subjects are provided in Sects. 6.5 and 6.6.

The microprocessor controls the entire system and enables real-time navigation by means of code pseudoranges. The control device provides interactive communication with the receiver. Commands can be keyed in and diagnostic or other messages can be displayed. Therefore, this device is usually designed as keyboard display unit.

A storage device is necessary for storing the observables and the navigation message so that they are available for later processing. Various media are presently used: microchips, cassette drives, magnetic bubbles or other nonvolatile storage. Additionally, the receiver can be interfaced to an external computer.

Many receivers have an (optional) internal power supply such as rechargeable batteries; but, there is always a provision for external batteries or other power supplies.

Radio frequency section. The radio frequency (RF) section forms the heart of the receiver. After signal input from the antenna, the discrimination of the signals is achieved by the C/A-codes for instance. These are, as mentioned previously, unique for each satellite. Codeless receivers must use other logic to discriminate between satellites. One technique is to monitor the Doppler shift. The RF section of single frequency units processes only the L_1 signal while dual frequency instruments process both the L_1 and L_2 signals. The data collected by a dual frequency receivers enable a combination where the ionospheric refraction can be eventually eliminated. This subject is treated in Sect. 6.3.2.

The incoming signals are usually assigned to a separate channel for each satellite. An important feature of the RF section is the number of channels and hence the number of satellites which can be tracked simultaneously. Older instruments used a limited number of physical channels and alternated satellite tracking by rapidly sequencing (20 milliseconds) satellites on the same channel. Today, most receivers assign one satellite each to a physical channel where the satellites are continuously tracked. Multichannel receivers are more accurate and less sensitive to loss of signal lock but show interchannel biases. However, with modern receivers these biases can be completely calibrated out at the level of less than 0.1 mm. Receivers with sequencing channels are less expensive but slower. The combination of both concepts can be found in hybrid receivers.

The basic elements of the RF section are oscillators to generate a reference frequency, filters to eliminate undesired frequencies, and mixers. In the latter two oscillations y_1, y_2 with different amplitudes a_1, a_2 and different frequencies f_1, f_2 are mathematically multiplied. In simplified form this gives

$$\begin{aligned} y = y_1\, y_2 &= a_1 \cos(f_1\, t)\, a_2 \cos(f_2\, t) \\ &= \frac{a_1\, a_2}{2} \left[\cos\left((f_1 - f_2)\, t\right) + \cos\left((f_1 + f_2)\, t\right) \right] \end{aligned} \tag{5.7}$$

and results in an oscillation y consisting of a low-frequency part and a high-frequency part. After applying a low-pass filter, the high-frequency part is eliminated. The remaining low-frequency signal is processed. The difference $f_1 - f_2$ between the frequencies is often called intermediate or beat frequency.

The actual phase measurement is performed in tracking loop circuits where two different techniques are used: (1) the code correlation technique requiring knowledge of one PRN code, and (2) codeless techniques which are independent of code. Both techniques reconstruct the unmodulated carrier wave from which the phase of the base carrier is measured.

The code correlation technique provides all components of the satellite signal: the satellite clock reading, the navigation message, and the unmodulated carrier. The procedure is performed in several steps. First, a reference carrier is generated in the receiver which then is biphase modulated with a replica of the known PRN code. In a second step, the resulting reference signal is correlated with the received satellite signal. The signals are shifted with respect to time so that they optimally match (based on a high mathematical correlation). The necessary time shift Δt, neglecting clock biases, corresponds to the travel time of the signal from the satellite antenna to the phase center of the receiving antenna. After removal of the PRN code, the received signal still contains the navigation message which can be decoded and eliminated by high-pass filtering. The final result is the Doppler shifted carrier on which a phase measurement can be performed. Since a PRN code is required, the code correlation technique is generally only applicable to the C/A-code with only the $L1$ carrier being reconstructed. If the P-code is available, both carriers can be reconstructed. Normally, the C/A-code is used to lock onto the signal and to initialize the tracking loop. This is performed in a short time since this code repeats every millisecond and is computationally simple. One result of the C/A-code correlation is the decoded navigation message which contains the hand-over word (HOW) in each subframe. The HOW tells the receiver where to start the search in the P-code for signal matching.

The codeless technique is based on signal squaring which is a procedure first presented in Counselman (1981). The received signal is mixed (i.e., multiplied) with itself and hence all modulations are removed. This happens because a 180° phase shift during modulation is equivalent to a change in the sign of the signal. The result is the unmodulated carrier with twice the frequency, cf. Eq. (5.7), and thus half the wavelength. Another codeless technique provides the phase of the PRN code as described by MacDoran et al. (1985). The codeless techniques have the advantage of being independent of PRN codes. A drawback of these techniques is the fact that the satellite clock and the satellite orbit information are unavailable. Therefore, no real-time navigation is possible, and the receivers must be time synchronized by external methods. Additionally, in case of squaring, the signal-to-noise ratio is substantially reduced since the noise is also squared.

In order to get both carriers without knowledge of the P-code, most of the receivers provide a hybrid technique. The $L1$ carrier is reconstructed by code correlation using the C/A-code, and squaring is applied to reconstruct the $L2$ carrier.

6. Observables

6.1 Data acquisition

In concept, the GPS observables are ranges which are deduced from measured time or phase differences based on a comparison between received signals and receiver generated signals. Unlike the terrestrial electronic distance measurements, GPS uses the "one way concept" where two clocks are used, namely one in the satellite and the other in the receiver. Thus, the ranges are biased by satellite and receiver clock errors and consequently they are denoted as pseudoranges.

6.1.1 Code pseudoranges

Let us denote t^S the reading of the satellite clock at emission time and t_R the reading of the receiver clock at signal reception time. Analogously, the delays of the clocks with respect to GPS system time will be termed δ^S and δ_R. Recall that the satellite clock reading t^S is transmitted via the PRN code. The difference between the clock readings is equivalent to the time shift Δt which aligns the satellite and reference signal during the code correlation procedure in the receiver. Thus,

$$\Delta t = t_R - t^S = [t_R(GPS) - \delta_R] - [t^S(GPS) - \delta^S] =$$
$$= \Delta t(GPS) + \Delta \delta, \tag{6.1}$$

with $\Delta t(GPS) = t_R(GPS) - t^S(GPS)$ and $\Delta \delta = \delta^S - \delta_R$. The bias δ^S of the satellite clock can be modeled by a polynomial with the coefficients being transmitted in the first subframe of the navigation message. Assuming the δ^S correction is applied, $\Delta \delta$ equals the negative receiver clock delay. The time interval Δt multiplied by the speed of light c yields the pseudorange R and hence

$$R = c\,\Delta t = c\,\Delta t(GPS) + c\,\Delta \delta = \varrho + c\,\Delta \delta. \tag{6.2}$$

The range ϱ is calculated from the true signal travel time. In other words, ϱ corresponds to the distance between the position of the satellite at epoch $t^S(GPS)$ and the position of receiver's antenna at epoch $t_R(GPS)$. Since ϱ

is a function of two different epochs, it is often expanded into a Taylor series
with respect to e.g. the emission time

$$\varrho = \varrho(t^S, t_R) = \varrho(t^S, (t^S + \Delta t))$$
$$= \varrho(t^S, t^S) + \dot{\varrho}(t^S, t^S)\, \Delta t \tag{6.3}$$

where $\dot{\varrho}$ denotes the time derivative of ϱ or the radial velocity of the satellite
relative to the receiving antenna. All epochs in Eq. (6.3) are expressed in
GPS system time.

The maximum radial velocity for GPS satellites in the case of a stationary
receiver is $\dot{\varrho} \approx 0.9\,\mathrm{km \cdot s^{-1}}$, cf. Sect. 5.1.1, and the travel time of the satellite
signal is about 0.07 seconds. The correction term in Eq. (6.3) thus amounts
to 60 m.

The precision of a pseudorange derived from code measurements has been
traditionally about 1% of the chip length. Thus, a precision of roughly 3 m
and 0.3 m is achieved with C/A-code and P-code pseudoranges, respectively.
However, recent developments indicate that a precision of about 0.1% of the
chip length may be possible.

6.1.2 Phase pseudoranges

Let us denote by $\varphi^S(t)$ the phase of the received and reconstructed carrier
with frequency f^S and by $\varphi_R(t)$ the phase of a reference carrier generated
in the receiver with frequency f_R. Here, the parameter t is an epoch in
the GPS time system reckoned from an initial epoch $t_0 = 0$. According to
Eq. (5.4) the following phase equations are obtained

$$\varphi^S(t) = f^S\, t - f^S\, \frac{\varrho}{c} - \varphi_0^S \tag{6.4}$$
$$\varphi_R(t) = f_R\, t - \varphi_{0R}.$$

The initial phases φ_0^S, φ_{0R} are caused by clock errors and are equal to

$$\varphi_0^S = f^S\, \delta^S \tag{6.5}$$
$$\varphi_{0R} = f_R\, \delta_R.$$

Hence, the beat phase $\varphi_R^S(t)$ is given by

$$\varphi_R^S(t) = \varphi^S(t) - \varphi_R(t)$$
$$= -f^S\, \frac{\varrho}{c} - f^S\, \delta^S + f_R\, \delta_R + (f^S - f_R)\, t. \tag{6.6}$$

The deviation of the frequencies f^S, f_R from the nominal frequency f is
only in the order of some fractional parts of Hz. This may be verified by

considering for instance a short time stability in the frequencies of $df/f = 10^{-12}$. With the nominal carrier frequency $f \approx 1.5\,\text{GHz}$, the frequency error thus becomes $df = 1.5 \cdot 10^{-3}\,\text{Hz}$. Such a frequency error can be neglected because during signal propagation (i.e., $t = 0.07$ seconds) a maximum error of 10^{-4} cycles in the beat phase is generated which is below the noise level. The clock errors are in the range of milliseconds and are thus yet less effective. Summarizing, Eq. (6.6) can be written in the more simple form

$$\varphi_R^S(t) = -f\,\frac{\varrho}{c} - f\,\Delta\delta\,. \tag{6.7}$$

where again $\Delta\delta = \delta^S - \delta_R$ has been introduced. If the assumption of frequency stability is incorrect and the oscillators are unstable, then their behaviour has to be modeled by e.g. polynomials where clock and frequency offsets and a frequency drift are determined. A complete carrier phase model which includes the solution of large (e.g., 1 second) receiver clock errors was developed by Remondi (1984). Adequate formulas can also be found in King et al. (1987), p. 55 ff. No further details are given here because in practice eventual residual errors will be eliminated by differencing the measurements.

Switching on a receiver at epoch t_0, the instantaneous fractional beat phase is measured. The initial integer number N of cycles between satellite and receiver is unknown. However, when tracking is continued without loss of lock, the number N, also called integer ambiguity, remains the same and the beat phase at epoch t is given by

$$\varphi_R^S(t) = \Delta\varphi_R^S\,\bigg|_{t_0}^{t} + N \tag{6.8}$$

where $\Delta\varphi_R^S$ denotes the (measurable) fractional phase at epoch t augmented by the number of integer cycles since the initial epoch t_0. A geometrical interpretation of Eq. (6.8) is provided in Fig. 6.1 where $\Delta\varphi_i$ is a shortened notation for $\Delta\varphi_R^S\,|_{t_0}^{t_i}$ and, for simplicity, the initial fractional beat phase $\Delta\varphi_0$ is assumed to be zero. Substituting Eq. (6.8) into Eq. (6.7) and denoting the negative observation quantity by $\Phi = -\Delta\varphi_R^S$ yields the equation for the phase pseudoranges

$$\Phi = \frac{1}{\lambda}\,\varrho + \frac{c}{\lambda}\,\Delta\delta + N \tag{6.9}$$

where the wavelength λ has been introduced according to Eq. (5.1). Multiplying the above equation by λ scales the phase expressed in cycles to a range which differs from the code pseudorange only by the integer multiples of λ. Again, the range ϱ represents the distance between the satellite at emission epoch t and the receiver at reception epoch $t + \Delta t$. The phase of

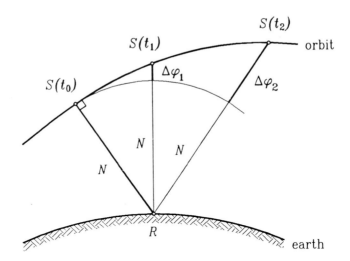

Fig. 6.1. Geometrical interpretation of phase range

the carrier can be measured to better than 0.01 cycles which corresponds to millimeter precision.

It should be noted that a plus sign convention has been chosen for Eq. (6.9). This choice is somehow arbitrary since quite often the phase Φ and the distance ϱ show different signs. Actually, the sign is receiver dependent because the beat phase is generated in the receiver and the combination of the satellite and the receiver signal differs for various receiver types.

6.1.3   Doppler data

Some of the first solution models proposed for GPS were to use the Doppler observable as with the TRANSIT system. This system used the integrated Doppler shifts (i.e., phase differences) which were scaled to delta ranges. The raw Doppler shift, cf. Eq. (5.5), being linearly dependent on the radial velocity and thus allowing for velocity determination in real-time is important for navigation. Considering Eq. (6.9), the equation for the observed Doppler shift scaled to range rate is given by

$$D = \lambda\,\dot{\Phi} = \dot{\varrho} + c\,\Delta\dot{\delta} \tag{6.10}$$

where the derivatives with respect to time are indicated by a dot. The raw Doppler shift is less accurate than integrated Doppler, cf. Hatch (1982). To get an idea of the achievable accuracy, Ashjaee et al. (1989) mention 0.001 Hz (this corresponds to $0.2\,\mathrm{mm\cdot s^{-1}}$).

A detailed derivation of Doppler equations within the frame of GPS is given in Remondi (1984) where even relativistic effects are accounted for. It is worth noting here that the raw Doppler shift is also applied to determine integer ambiguities in kinematic surveying, cf. Remondi (1991a), or is used as an additional independent observable for point positioning, cf. Ashjaee et al. (1989).

6.1.4 Biases and noise

The code pseudoranges, cf. Eq. (6.2), and phase pseudoranges, cf. Eq. (6.9), are affected by both, systematic errors or biases and random noise. Note that Doppler measurements are affected by the rate of change of the biases only. The error sources can be classified into three groups, namely satellite related errors, propagation medium related errors, and receiver related errors. Some range biases are listed in Table 6.1.

The systematic errors can be modeled and give rise to additional terms in the observation equations which will be explained in detail in subsequent sections. As mentioned earlier, systematic effects can also be eliminated by appropriate combinations of the observables. Differencing between receivers eliminates satellite-specific biases, and differencing between satellites eliminates receiver-specific biases. Thus, double-differenced pseudoranges are, to a high degree, free of systematic errors originating from the satellites and from the receivers. With respect to refraction, this is only true for short baselines where the measured ranges at both endpoints are affected equally. In addition, ionospheric refraction can be virtually eliminated by an adequate combination of dual frequency data.

The random noise mainly contains the actual observation noise plus multipath effects caused by multiple reflections of the signal (which can also

Table 6.1. Range biases

Source	Effect
Satellite	Orbital errors, Clock bias
Signal propagation	Tropospheric refraction, Ionospheric refraction
Receiver	Antenna phase center variation, Clock bias

occur at the satellite during signal emission). Multipath is interference between the direct and the reflected signal and is largely random; however, it
may also appear as a short term bias. Wells et al. (1987) report a similar
effect called "imaging" where a reflecting obstacle generates an image of the
real antenna which distorts the antenna pattern. Both effects, multipath
and imaging, can be considerably reduced by selecting sites protected from
reflections (buildings, vehicles, trees, etc.) and by an appropriate antenna
design. It should be noted that multipath is frequency dependent. Therefore, carrier phases are less affected than code ranges where multipath can
amount to the meter level as stated by Lachapelle (1990). More details on
the multipath problems are given in Sect. 6.5.

The measurement noise, an estimation of the satellite biases, and the
contributions from the wave propagation are combined in the User Equivalent Range Error (UERE). This UERE is transmitted via the navigation
message. In combination with a DOP factor, explained in Sect. 9.5, UERE
allows for an estimation of the achievable point positioning precision.

6.2   Data combinations

GPS observables are obtained from the code information or the carrier wave
in the broadcast satellite signal. Recall that the P-code is modulated on both
carriers $L1$ and $L2$ whereas the C/A-code is modulated on $L1$ only. Consequently, one could measure the carrier phases Φ_{L1}, Φ_{L2}, the corresponding
Doppler shifts D_{L1}, D_{L2}, and the code ranges $R_{L1,C/A}$, $R_{L1,P}$, $R_{L2,P}$ for a
single epoch. In reality, not all observables are available because, for example, a single frequency receiver delivers only data from the $L1$ frequency.
Furthermore, the Doppler observables are not considered here.

The objectives of this section are to show how linear combinations are
developed for dual frequency phases, and how code range smoothing by
means of carrier phases is performed.

6.2.1   Linear phase combinations

Generally, the linear combination of two phases φ_1 and φ_2 is defined by

$$\varphi = n_1 \varphi_1 + n_2 \varphi_2 \tag{6.11}$$

where n_1 and n_2 are arbitrary numbers. Substitution of the relations $\varphi_i = f_i t$ for the corresponding frequencies f_1 and f_2 yields

$$\varphi = n_1 f_1 t + n_2 f_2 t = f t. \tag{6.12}$$

Therefore,

$$f = n_1 \, f_1 + n_2 \, f_2 \tag{6.13}$$

is the frequency and

$$\lambda = \frac{c}{f} \tag{6.14}$$

is the wavelength of the linear combination.

In the case of GPS, the linear combination of $L1$ and $L2$ carrier phases Φ_{L1} and Φ_{L2} for the simplest nontrivial cases in Eq. (6.11) are $n_1 = n_2 = 1$, yielding the sum

$$\Phi_{L1+L2} = \Phi_{L1} + \Phi_{L2} \tag{6.15}$$

and $n_1 = 1$, $n_2 = -1$, leading to the difference

$$\Phi_{L1-L2} = \Phi_{L1} - \Phi_{L2} \, . \tag{6.16}$$

The corresponding wavelengths according to Eq. (6.14) are

$$\lambda_{L1+L2} = 10.7 \text{ cm}$$
$$\tag{6.17}$$
$$\lambda_{L1-L2} = 86.2 \text{ cm}$$

where the numerical values for the carrier frequencies f_{L1} and f_{L2}, see Table 5.2, p. 71, have been substituted. The combination Φ_{L1+L2} is denoted as narrow lane and Φ_{L1-L2} as wide lane, cf. for example Beutler et al. (1988), Wübbena (1988).

A slightly more complicated linear combination results from the choice

$$n_1 = 1 \qquad n_2 = -\frac{f_{L2}}{f_{L1}} \tag{6.18}$$

which is frequently denoted as $L3$ signal, thus

$$\Phi_{L3} = \Phi_{L1} - \frac{f_{L2}}{f_{L1}} \, \Phi_{L2} \, . \tag{6.19}$$

Now that the significant linear combinations have been defined, the advantage of these combinations will be shown. The $L3$ combination, for example, is used to reduce ionospheric effects, cf. Sect. 6.3.2., and the ambiguity resolution combines wide and narrow lane signals, cf. Sect. 9.1.3.

Assuming a certain noise level for the phase measurements, it is seen that the noise level will increase for these linear combinations. Applying the error propagation law and assuming the same noise for both phases, the noise of the sum or the difference formed by Φ_{L1} and Φ_{L2} is higher by the factor $\sqrt{2}$ than the noise of a single phase. Of course, to compute this factor correctly, one must take into account the different noise levels.

6.2.2   Phase and code pseudorange combinations

The objective here is to show the principle of the smoothing of code pseudoranges by means of phase pseudoranges. A first extensive investigation
of this subject was provided by Hatch (1982). Applications and improvements were proposed later by the same author and are given in Hatch and
Larson (1985) and Hatch (1986). Other slight variations can be found in
e.g. Lachapelle et al. (1986) or Meyerhoff and Evans (1986). Today, phase
and code pseudorange combinations are an important part of real-time trajectory determination.

Assuming dual frequency measurements for epoch t_1, the P-code pseudo-
ranges $R_{L1}(t_1)$, $R_{L2}(t_1)$ and the carrier phase pseudoranges $\Phi_{L1}(t_1)$, $\Phi_{L2}(t_1)$
are obtained. Assume also, the code pseudoranges are scaled to cycles (but
still denoted as R) by dividing them by the corresponding carrier wavelength.
Using the two frequencies f_{L1}, f_{L2}, the combination

$$R(t_1) = \frac{f_{L1}\, R_{L1}(t_1) - f_{L2}\, R_{L2}(t_1)}{f_{L1} + f_{L2}} \tag{6.20}$$

is formed for the code pseudoranges and the wide lane signal

$$\Phi(t_1) = \Phi_{L1}(t_1) - \Phi_{L2}(t_1) \tag{6.21}$$

for the carrier phase pseudoranges. From Eq. (6.20) one can see that the
noise of the combined code pseudorange $R(t_1)$ is reduced by a factor of 0.7
compared to the noise of the single code measurement. The increase of the
noise in the wide lane signal by a factor of $\sqrt{2}$ has no effect because the
noise of the carrier phase pseudoranges is essentially lower than the noise
of the code pseudoranges. It is worth noting that both signals $R(t_1)$ and
$\Phi(t_1)$ have the same frequency and thus the same wavelength as the reader
may verify by applying Eq. (6.13). This is not the case with the approach
given in Hatch (1986) where a plus sign is chosen for the code pseudorange
combination (this can be regarded as a weighted mean).

Combinations of the form (6.20) and (6.21) are formed for each epoch.
Additionally, for all epochs t_i after t_1, extrapolated values of the code pseudoranges $R(t_i)_{\text{ex}}$ can be calculated from

$$R(t_i)_{\text{ex}} = R(t_1) + (\Phi(t_i) - \Phi(t_1)). \tag{6.22}$$

The smoothed value $R(t_i)_{\text{sm}}$ is finally obtained by the arithmetic mean

$$R(t_i)_{\text{sm}} = \frac{1}{2}\left(R(t_i) + R(t_i)_{\text{ex}}\right). \tag{6.23}$$

Generalizing the above formulas for an arbitrary epoch t_i (with the preceding epoch t_{i-1}), a recursive algorithm is given by

$$R(t_i) = \frac{f_{L1} R_{L1}(t_i) - f_{L2} R_{L2}(t_i)}{f_{L1} + f_{L2}} ,$$

$$\Phi(t_i) = \Phi_{L1}(t_i) - \Phi_{L2}(t_i),$$

$$R(t_i)_{ex} = R(t_{i-1})_{sm} + (\Phi(t_i) - \Phi(t_{i-1})),$$

$$R(t_i)_{sm} = \frac{1}{2} (R(t_i) + R(t_i)_{ex})$$

which works for all $i > 1$ under the initial condition $R(t_1) = R(t_1)_{ex} = R(t_1)_{sm}$.

The above algorithm assumes the data is free of gross errors. However, carrier phase data are sensitive to changes in the integer ambiguity (i.e., cycle slips). To circumvent this problem, a variation of the algorithm is given in Lachapelle et al. (1986). Using the same notations as before for an epoch t_i, the smoothed code pseudorange is obtained by

$$R(t_i)_{sm} = w R(t_i) + (1 - w)(R(t_{i-1})_{sm} + \Phi(t_i) - \Phi(t_{i-1})) \quad (6.24)$$

with w as a time dependent weight factor. For the first epoch $i = 1$, the weight is set $w = 1$, thus putting the full weight on the measured code pseudorange. For consecutive epochs, the weight is reduced continuously and thus emphasizes the influence of the carrier phases. To get an idea for the reduction factor, Lachapelle et al. (1986) proposed a reduction of the weight by 0.01 from epoch to epoch for a kinematic experiment with a data sampling rate of 1.2 seconds. After two minutes, only the smoothed value of the previous epoch (augmented by the carrier phase difference) is taken into account. Again, in case of cycle slips, the algorithm would fail. A simple check of the carrier phase difference for two consecutive epochs by the Doppler shift × time may detect data irregularities such as cycle slips. After the occurrence of a cycle slip, the weight is reset to $w = 1$ which fully eliminates the influence of the erroneous carrier phase data. The key of this approach is that cycle slips must be detected but do not have to be corrected, cf. Hein et al. (1988).

Another smoothing algorithm for code pseudoranges is given by Meyerhoff and Evans (1986). Here, the phase changes obtained for instance by the integrated Doppler shift between the epochs t_i and t_1 are denoted by $\Delta\Phi(t_i, t_1)$ with t_1 being the starting epoch for the integration. Accordingly from each code pseudorange $R(t_i)$ at epoch t_i, an estimate of the code pseudorange at epoch t_1 can be given by

$$R(t_1)_i = R(t_i) - \Delta\Phi(t_i, t_1) \quad (6.25)$$

where the subscript i on the left-hand side indicates the epoch that the code pseudorange $R(t_1)$ is computed from. Obtaining consecutively for each epoch an estimate, the arithmetic mean $R(t_1)$m of the code pseudorange for n epochs is calculated by

$$R(t_1)\text{m} = \frac{1}{n} \sum_{i=1}^{n} R(t_1)_i \qquad (6.26)$$

and the smoothed code pseudorange for an arbitrary epoch results from

$$R(t_i)\text{sm} = R(t_1)\text{m} + \Delta\Phi(t_i, t_1). \qquad (6.27)$$

The advantage of this procedure lies in the reduction of the noise in the initial code pseudorange by averaging an arbitrary number n of measured code pseudoranges. Note from the three formulas (6.25) through (6.27) that the algorithm may also be applied successively epoch by epoch where the arithmetic mean must be updated from epoch to epoch. Using the above notations, formula (6.27) also works for epoch t_1, where of course $\Delta\Phi(t_1, t_1)$ is zero and there is no smoothing effect.

All the smoothing algorithms are also applicable if only single frequency data are available. In this case $R(t_i)$, $\Phi(t_i)$, and $\Delta\Phi(t_i, t_1)$ denote the single frequency code pseudorange, carrier phase pseudorange, and phase difference, respectively.

6.3   Atmospheric effects

6.3.1   Phase and group velocity

Consider a single electromagnetic wave propagating in space with wavelength λ and frequency f. The velocity of its phase

$$v_{ph} = \lambda f \qquad (6.28)$$

is denoted phase velocity. For GPS, the carrier waves $L1$ and $L2$ are propagating with this velocity.

For a group of waves with slightly different frequencies, the propagation of the resultant energy is defined by the group velocity

$$v_{gr} = -\frac{df}{d\lambda} \lambda^2 \qquad (6.29)$$

according to e.g. Bauer (1989), p. 110. This velocity has to be considered for GPS code measurements.

A relation between phase and group velocity may be derived by forming the total differential of Eq. (6.28) resulting in

$$dv_{ph} = f\,d\lambda + \lambda\,df \tag{6.30}$$

which can be rearranged to

$$\frac{df}{d\lambda} = \frac{1}{\lambda}\frac{dv_{ph}}{d\lambda} - \frac{f}{\lambda}. \tag{6.31}$$

Substitution of (6.31) into (6.29) yields

$$v_{gr} = -\lambda\frac{dv_{ph}}{d\lambda} + f\,\lambda \tag{6.32}$$

or finally the Rayleigh equation

$$v_{gr} = v_{ph} - \lambda\frac{dv_{ph}}{d\lambda}. \tag{6.33}$$

Note that the differentiation (6.30) implicitly contains the dispersion, cf. Joos (1956), p. 57, which is defined as a dependence of the phase velocity on the wavelength or the frequency. Phase and group velocity are equal in nondispersive media and correspond to the speed of light $c = 299\,792\,458\,\mathrm{m\cdot s^{-1}}$ in vacuum.

The wave propagation in a medium depends on the refractive index n. Generally, the propagation velocity is obtained from

$$v = \frac{c}{n}. \tag{6.34}$$

Applying this expression to the phase and group velocity, one gets appropriate formulas for the corresponding refractive indices n_{ph} and n_{gr}:

$$v_{ph} = \frac{c}{n_{ph}} \tag{6.35}$$

$$v_{gr} = \frac{c}{n_{gr}}. \tag{6.36}$$

Differentiation of the phase velocity with respect to λ, that is

$$\frac{dv_{ph}}{d\lambda} = -\frac{c}{n_{ph}^2}\frac{dn_{ph}}{d\lambda}, \tag{6.37}$$

and substitution of the last three equations into (6.33) yields

$$\frac{c}{n_{gr}} = \frac{c}{n_{ph}} + \lambda\frac{c}{n_{ph}^2}\frac{dn_{ph}}{d\lambda} \tag{6.38}$$

or

$$\frac{1}{n_{gr}} = \frac{1}{n_{ph}} \left(1 + \lambda \frac{1}{n_{ph}} \frac{dn_{ph}}{d\lambda} \right) . \tag{6.39}$$

This equation may be inverted to

$$n_{gr} = n_{ph} \left(1 - \lambda \frac{1}{n_{ph}} \frac{dn_{ph}}{d\lambda} \right) \tag{6.40}$$

where the approximation $(1 + \varepsilon)^{-1} = 1 - \varepsilon$ has been applied accordingly. Thus,

$$n_{gr} = n_{ph} - \lambda \frac{dn_{ph}}{d\lambda} \tag{6.41}$$

is the modified Rayleigh equation. A slightly different form is obtained by differentiating the relation $c = \lambda f$ with respect to λ and f, that is

$$\frac{d\lambda}{\lambda} = -\frac{df}{f} , \tag{6.42}$$

and by substituting the result into (6.41):

$$n_{gr} = n_{ph} + f \frac{dn_{ph}}{df} . \tag{6.43}$$

6.3.2 Ionospheric refraction

The ionosphere, extending in various layers from about $50\,km$ to $1\,000\,km$ above earth, is a dispersive medium with respect to the GPS radio signal. Following Seeber (1989), p. 50, the series

$$n_{ph} = 1 + \frac{c_2}{f^2} + \frac{c_3}{f^3} + \frac{c_4}{f^4} + \cdots \tag{6.44}$$

approximates the phase refractive index. The coefficients c_2, c_3, c_4 do not depend on frequency but on the quantity N_e denoting the number of electrons per m^3 (i.e., the electron density) along the propagation path. Using an approximation by cutting off the series expansion after the quadratic term, that is

$$n_{ph} = 1 + \frac{c_2}{f^2} , \tag{6.45}$$

differentiating this equation

$$dn_{ph} = -\frac{2c_2}{f^3} df , \tag{6.46}$$

and substituting (6.45) and (6.46) into (6.43) yields

$$n_{gr} = 1 + \frac{c_2}{f^2} - f\,\frac{2c_2}{f^3} \tag{6.47}$$

or

$$n_{gr} = 1 - \frac{c_2}{f^2}\,. \tag{6.48}$$

It can be seen from (6.45) and (6.48) that the group and the phase refractive indices deviate from unity with opposite sign. Introducing an estimate for c_2, cf. Seeber (1989), p. 50, by

$$c_2 = -40.3\ N_e\quad [\text{Hz}^2] \tag{6.49}$$

then $n_{gr} > n_{ph}$ and thus $v_{gr} < v_{ph}$ follows. As a consequence of the different velocities, a group delay and a phase advance occurs, cf. for example Young et al. (1985). In other words, GPS code measurements are delayed and the carrier phases are advanced. Therefore, the code pseudoranges are measured too long and the carrier phase pseudoranges are measured too short compared to the geometric distance between the satellite and the receiver. The amount of the difference is in both cases the same.

According to Fermat's principle, the measured range s is defined by

$$s = \int n\,ds \tag{6.50}$$

where the integral must be extended along the path of the signal. The geometric distance s_0 is measured along the straight line between the satellite and the receiver and thus may be obtained analogously by setting $n = 1$:

$$s_0 = \int ds_0\,. \tag{6.51}$$

The difference Δ^{Iono} between measured and geometric range is called ionospheric refraction and follows from

$$\Delta^{Iono} = \int n\,ds - \int ds_0 \tag{6.52}$$

which may be written for a phase refractive index n_{ph} from (6.45) as

$$\Delta^{Iono}_{ph} = \int \left(1 + \frac{c_2}{f^2}\right) ds - \int ds_0 \tag{6.53}$$

and for a group refractive index n_{gr} from (6.48) as

$$\Delta^{Iono}_{gr} = \int \left(1 - \frac{c_2}{f^2}\right) ds - \int ds_0\,. \tag{6.54}$$

A simplification is obtained when allowing the integration for the first term
in (6.53) and (6.54) along the geometric path. In this case ds becomes ds_0
and the formulas

$$\Delta_{ph}^{Iono} = \int \frac{c_2}{f^2} \, ds_0 \qquad \Delta_{gr}^{Iono} = -\int \frac{c_2}{f^2} \, ds_0 \qquad (6.55)$$

result which can also be written as

$$\Delta_{ph}^{Iono} = -\frac{40.3}{f^2} \int N_e \, ds_0 \qquad \Delta_{gr}^{Iono} = \frac{40.3}{f^2} \int N_e \, ds_0 \qquad (6.56)$$

where (6.49) has been substituted. Defining the total electron content by

$$\text{TEC} = \int N_e \, ds_0, \qquad (6.57)$$

and substituting TEC into (6.56) yields

$$\Delta_{ph}^{Iono} = -\frac{40.3}{f^2} \text{TEC} \qquad \Delta_{gr}^{Iono} = \frac{40.3}{f^2} \text{TEC} \qquad (6.58)$$

as the final result which has the dimension of length. Usually, the TEC is
measured in units of 10^{16} electrons per m^2.

Because the integration in Eq. (6.57) has been performed along the vertical direction, formula (6.58) suffices for a satellite at zenith. For arbitrary
lines of sight, the zenith distance of the satellite must be taken into account
by

$$\Delta_{ph}^{Iono} = -\frac{1}{\cos z'} \frac{40.3}{f^2} \text{TEC} \qquad \Delta_{gr}^{Iono} = \frac{1}{\cos z'} \frac{40.3}{f^2} \text{TEC} \qquad (6.59)$$

since the path length in the ionosphere varies with a changing zenith distance.

From Fig. 6.2, illustrating the situation, the relation

$$\sin z' = \frac{R_E}{R_E + h_m} \sin z \qquad (6.60)$$

can be read where R_E is the mean radius of the earth, h_m is a mean value
for the height of the ionosphere, and z' and z are the zenith distances at the
ionospheric point (IP) and at the observing site. The zenith distance z can
be calculated for a known satellite position and approximate coordinates of
the observation location. For h_m a value in the range between 300 km and
400 km is typical. Gervaise et al. (1985) use 300 km, Wild et al. (1989) take
350 km, and Finn and Matthewman (1989) recommend an average value
of 400 km but prefer an algorithm for calculating individual mean heights.
Anyway, the height is only sensitive for low satellite elevations.

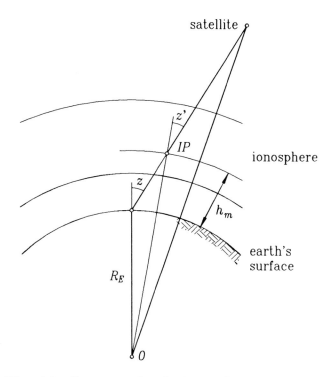

Fig. 6.2. Geometry for the ionospheric path delay

As shown by (6.58), the change of range caused by the ionospheric refraction may be restricted to the determination of the total electron content (TEC). However, the TEC itself is a fairly complicated quantity because it depends on sunspot activities (approximately 11-year cycle), seasonal and diurnal variations, the line of sight which includes elevation and azimuth of the satellite, and the position of the observation site, cf. e.g. Finn and Matthewman (1989). Taking all of these effects into account, a GPS pseudorange may be wrong from about 0.15 m to 50 m, cf. Clynch and Coco (1986). The TEC may be measured, estimated, its effect computed by models, or eliminated.

Measuring the TEC. Considering as an example Japan, Kato et al. (1987) describe one facility in Tokyo that directly measures the TEC. However, since there is a correlation between the TEC and the critical plasma frequency, the TEC may also be calculated by the five Japanese ionospheric observatories which make available critical plasma frequency results on an hourly basis. Using the calculated instead of the measured TEC induces an error of about 20%, but by interpolation any arbitrary location in Japan can be covered.

Estimating the TEC. A straightforward estimation of the TEC is described by Wild et al. (1989) where a Taylor series expansion as function of the observation latitude and the local solar time is set up and substituted into (6.59). The coefficients in the Taylor series are introduced as unknowns in the pseudorange equations and estimated together with the other unknowns during data processing.

Computing the effect of the TEC. Here, the entire vertical ionospheric refraction is approximated by the Klobuchar (1986) model and yields the vertical time delay for the code measurements. Although the model is an approximation, it is nevertheless of importance because it uses the ionospheric coefficients broadcast within the fourth subframe of the navigation message, cf. Sect. 5.1.2. Following Jorgensen (1989), the Klobuchar model is

$$\Delta T_v^{Iono} = A_1 + A_2 \cos\left(\frac{2\pi(t - A_3)}{A_4}\right) \tag{6.61}$$

where

$$A_1 = 5 \cdot 10^{-9} \text{ s} = 5 \text{ ns}$$

$$A_2 = \alpha_1 + \alpha_2 \, \varphi_{IP}^m + \alpha_3 \, \varphi_{IP}^m{}^2 + \alpha_4 \, \varphi_{IP}^m{}^3$$

$$A_3 = 14^h \text{ local time} \tag{6.62}$$

$$A_4 = \beta_1 + \beta_2 \, \varphi_{IP}^m + \beta_3 \, \varphi_{IP}^m{}^2 + \beta_4 \, \varphi_{IP}^m{}^3 \,.$$

The values for A_1 and A_3 are constant, the coefficients α_i, β_i, $i = 1, \ldots, 4$ are uploaded daily to the satellites and broadcast to the user. The parameter t in (6.61) is the local time of the ionospheric point IP, cf. Fig. 6.2, and may be derived from

$$t = \frac{\lambda_{IP}}{15} + t_{UT} \tag{6.63}$$

where λ_{IP} is the geomagnetic longitude positive to East for the ionospheric point in degree and t_{UT} is the observation epoch in universal time, cf. Walser (1988). Finally, φ_{IP}^m in Eq. (6.62) is the spherical distance between the geomagnetic pole and the ionospheric point. Denoting the coordinates of the geomagnetic pole by φ_P, λ_P and those of the ionospheric point by φ_{IP}, λ_{IP}, then $\sin \varphi_{IP}^m$ is obtained by

$$\sin \varphi_{IP}^m = \sin \varphi_{IP} \sin \varphi_P + \cos \varphi_{IP} \cos \varphi_P \cos(\lambda_{IP} - \lambda_P) \tag{6.64}$$

where at present, cf. Walser (1988), the coordinates of the geomagnetic pole are

$$\varphi_P = \ 78\overset{\circ}{.}3$$

$$\lambda_P = 291\overset{\circ}{.}0 \,. \tag{6.65}$$

Summarizing, the evaluation of the Klobuchar model may be performed by the following steps, cf. also Jorgensen (1989):

- For epoch t_{UT} compute the azimuth a and the zenith distance z of the satellite.

- Choose a mean height of the ionosphere and compute the distance s between the observing site and the ionospheric point obtained from the triangle origin-observation site-IP, cf. Fig. 6.2.

- Compute the coordinates φ_{IP}, λ_{IP} of the ionospheric point by means of the quantities a, z, s.

- Calculate φ_{IP}^m from (6.64).

- Calculate A_2 and A_4 from (6.62) where the coefficients α_i, β_i, $i = 1, \ldots, 4$ are received via the satellite navigation message.

- Use (6.62) and (6.63) and compute the vertical delay ΔT_v^{Iono} by (6.61).

- By calculating z' from (6.60) and applying $\Delta T^{Iono} = \frac{1}{\cos z'} \Delta T_v^{Iono}$, the transition from the vertical delay to the delay along the wave path is achieved. The result is obtained as a time delay in seconds which must be multiplied by the speed of light to get it as a change of range.

Eliminating the effect of the TEC. It is difficult to find a satisfying model for the TEC because of the various time dependent influences. The most efficient method, thus, is to eliminate the ionospheric refraction by using two signals with different frequencies. This dual frequency method is the main reason why the GPS signal has two carrier waves $L1$ and $L2$.

Starting with the code pseudorange model (6.2) and adding the frequency dependent ionospheric refraction, gives

$$R_{L1} = \varrho + c\,\Delta\delta + \Delta^{Iono}(f_{L1})$$
$$R_{L2} = \varrho + c\,\Delta\delta + \Delta^{Iono}(f_{L2}) \tag{6.66}$$

for the code ranges R_{L1} and R_{L2}. The frequencies of the two carriers are denoted by f_{L1} and f_{L2}, and the ionospheric term is equivalent to the group delay in Eq. (6.59).

A linear combination is now formed by

$$R_{L1,L2} = n_1\,R_{L1} + n_2\,R_{L2} \tag{6.67}$$

where n_1 and n_2 are arbitrary factors to be determined. The objective is to find a combination so that the ionospheric refraction cancels out. Substituting (6.66) into (6.67) leads to the postulate

$$n_1 \, \Delta^{Iono}(f_{L1}) + n_2 \, \Delta^{Iono}(f_{L2}) \overset{!}{=} 0 \qquad (6.68)$$

where the exclamation mark stresses that the expression must become zero. Equation (6.68) comprises two unknowns; therefore, one unknown may be chosen arbitrarily. Assuming $n_1 = 1$,

$$n_2 = -\frac{\Delta^{Iono}(f_{L1})}{\Delta^{Iono}(f_{L2})} \qquad (6.69)$$

follows or, by using (6.58), this may be written as

$$n_2 = -\frac{f_{L2}^2}{f_{L1}^2} \, . \qquad (6.70)$$

Substituting these values for n_1 and n_2, Eq. (6.68) is fulfilled and the linear combination (6.67) becomes

$$R_{L1,L2} = R_{L1} - \frac{f_{L2}^2}{f_{L1}^2} \, R_{L2} \, . \qquad (6.71)$$

This is the ionospheric-free linear combination for code ranges. A similar ionospheric-free linear combination for carrier phases may be derived. The carrier phase models can be written as

$$\lambda_{L1}\Phi_{L1} = \varrho + c \, \Delta\delta + \lambda_{L1} N_{L1} - \Delta^{Iono}(f_{L1})$$
$$\lambda_{L2}\Phi_{L2} = \varrho + c \, \Delta\delta + \lambda_{L2} N_{L2} - \Delta^{Iono}(f_{L2}) \qquad (6.72)$$

or, divided by the corresponding wavelengths,

$$\Phi_{L1} = \frac{1}{\lambda_{L1}} \varrho + f_{L1} \, \Delta\delta + N_{L1} - \frac{1}{\lambda_{L1}} \Delta^{Iono}(f_{L1})$$
$$\Phi_{L2} = \frac{1}{\lambda_{L2}} \varrho + f_{L2} \, \Delta\delta + N_{L2} - \frac{1}{\lambda_{L2}} \Delta^{Iono}(f_{L2}) \qquad (6.73)$$

which is linearly combined by

$$\Phi_{L1,L2} = n_1 \, \Phi_{L1} + n_2 \, \Phi_{L2} \qquad (6.74)$$

or explicitly

$$\Phi_{L1,L2} = \varrho \left(\frac{n_1}{\lambda_{L1}} + \frac{n_2}{\lambda_{L2}} \right) + \Delta\delta(n_1 f_{L1} + n_2 f_{L2})$$
$$+ n_1 N_{L1} + n_2 N_{L2} \tag{6.75}$$
$$- \frac{n_1}{\lambda_{L1}} \Delta^{Iono}(f_{L1}) - \frac{n_2}{\lambda_{L2}} \Delta^{Iono}(f_{L2}).$$

For an ionospheric-free linear phase combination the postulate

$$\frac{n_1}{\lambda_{L1}} \Delta^{Iono}(f_{L1}) + \frac{n_2}{\lambda_{L2}} \Delta^{Iono}(f_{L2}) \overset{!}{=} 0 \tag{6.76}$$

must be fulfilled where again one of the two unknowns n_1, n_2 may be arbitrarily chosen, thus

$$n_1 = 1$$
$$\tag{6.77}$$
$$n_2 = -\frac{\lambda_{L2}}{\lambda_{L1}} \frac{\Delta^{Iono}(f_{L1})}{\Delta^{Iono}(f_{L2})}$$

is a possible solution. Substituting again Eq. (6.58) and the relation $c = \lambda f$, one gets

$$n_1 = 1$$
$$\tag{6.78}$$
$$n_2 = -\frac{f_{L2}}{f_{L1}},$$

and the ionospheric-free linear phase combination is

$$\Phi_{L1,L2} = \Phi_{L1} - \frac{f_{L2}}{f_{L1}} \Phi_{L2}. \tag{6.79}$$

This result corresponds to Eq. (6.19). Note that the choice of this linear combination is somewhat arbitrary since $n_1 = 1$ was taken. For another choice see e.g. Beutler et al. (1988). However, since the noise of the ionospheric-free combination is increased compared to the raw phase, the choice of one of the two unknowns is restricted.

The elimination of the ionospheric refraction is the huge advantage of the two ionospheric-free linear combinations (6.71) and (6.79). Remembering the derivation, it should be clear that the term "ionospheric-free" is not fully correct because there are some approximations involved, for instance Eq. (6.45), the integration not along the true signal path (6.55), etc.

In the case of carrier phases, the ionospheric-free linear combination also has a significant disadvantage: if the ambiguities N_{L1} and N_{L2} in (6.73) are assumed to be integer numbers, then the linear combination gives a number $N = n_1 N_{L1} + n_2 N_{L2} = N_{L1} - (f_{L2}/f_{L1}) N_{L2}$ which is no longer an integer.

6.3.3   Tropospheric refraction

The effect of the neutral atmosphere (i.e., the nonionized part) is denoted
as tropospheric refraction, tropospheric path delay or simply tropospheric
delay. As Elgered et al. (1985) mention, the notations are slightly incorrect because they hide the stratosphere which is another constituent of the
neutral atmosphere. However, the dominant contribution of the troposphere
explains the notation.

The neutral atmosphere is a nondispersive medium with respect to radio
waves up to frequencies of 15 GHz, cf. for example Bauersima (1983), and
thus the propagation is frequency independent. Consequently, a distinction
between carrier phases and code ranges derived from different carriers $L1$ or
$L2$ is not necessary. The disadvantage is that an elimination of the tropospheric refraction by dual frequency methods is not possible.

The tropospheric path delay is defined by

$$\Delta^{Trop} = \int (n - 1)\, ds \tag{6.80}$$

which is analogous to the ionospheric formula (6.52). Again an approximation is introduced so that the integration is performed along the geometric
path of the signal. Usually, instead of the refractive index n the refractivity

$$N^{Trop} = 10^6\,(n - 1) \tag{6.81}$$

is used so that Eq. (6.80) becomes

$$\Delta^{Trop} = 10^{-6} \int N^{Trop}\, ds\,. \tag{6.82}$$

Hopfield (1969) shows the possibility of separating N^{Trop} into a dry and a
wet component

$$N^{Trop} = N_d^{Trop} + N_w^{Trop} \tag{6.83}$$

where the dry part results from the dry atmosphere and the wet part from
the water vapor. Correspondingly, the relations

$$\Delta_d^{Trop} = 10^{-6} \int N_d^{Trop}\, ds \tag{6.84}$$

$$\Delta_w^{Trop} = 10^{-6} \int N_w^{Trop}\, ds \tag{6.85}$$

and

$$\Delta^{Trop} = \Delta_d^{Trop} + \Delta_w^{Trop}$$

$$= 10^{-6} \int N_d^{Trop} \, ds + 10^{-6} \int N_w^{Trop} \, ds \tag{6.86}$$

are obtained. About 90% of the tropospheric refraction arise from the dry and about 10% from the wet component, cf. Janes et al. (1989). In practice, models for the refractivities are introduced in Eq. (6.86) and the integration is performed by numerical methods or analytically after e.g. series expansions of the integrand. Models for the dry and wet refractivity at the earth's surface have been known for some time, cf. for example Essen and Froome (1951). The corresponding dry component is

$$N_{d,0}^{Trop} = \bar{c}_1 \frac{p}{T}, \qquad \bar{c}_1 = 77.64 \left[\frac{K}{mb}\right] \tag{6.87}$$

where p is the atmospheric pressure in millibars (mb) and T is the temperature in Kelvin (K). The wet component was found to be

$$N_{w,0}^{Trop} = \bar{c}_2 \frac{e}{T} + \bar{c}_3 \frac{e}{T^2} \qquad \bar{c}_2 = -12.96 \left[\frac{K}{mb}\right]$$

$$\bar{c}_3 = 3.718 \cdot 10^5 \left[\frac{K^2}{mb}\right] \tag{6.88}$$

where e is the partial pressure of water vapor in mb and T again the temperature in K. The overbar in the coefficients only stresses that there is absolutely no relationship to the coefficients for the ionosphere in e.g. (6.55).

The values for $\bar{c}_1$, $\bar{c}_2$, and $\bar{c}_3$ are empirically determined and, certainly, cannot fully describe the local situation. An improvement is obtained by measuring meteorological data at the observation site. The following paragraphs present several models where meteorological surface data are taken into account.

Hopfield model. Using real data covering the whole earth, Hopfield (1969) has found empirically a representation of the dry refractivity as a function of the height h above the surface by

$$N_d^{Trop}(h) = N_{d,0}^{Trop} \left[\frac{h_d - h}{h_d}\right]^4 \tag{6.89}$$

under the assumption of a single polytropic layer with thickness

$$h_d = 40\,136 + 148.72\,(T - 273.16) \quad [m], \tag{6.90}$$

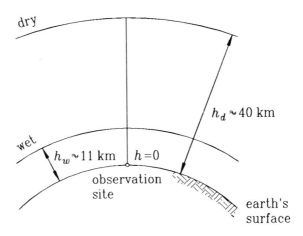

Fig. 6.3. Thickness of polytropic layers for the troposphere

see Janes et al. (1989) and cf. Fig. 6.3. Substitution of (6.89) and (6.90) into (6.84) yields (for the dry part) the tropospheric path delay

$$\Delta_d^{Trop} = 10^{-6} \, N_{d,0}^{Trop} \int \left[\frac{h_d - h}{h_d} \right]^4 ds \,. \tag{6.91}$$

The integral can be solved if the delay is calculated along the vertical direction and if the curvature of the signal path is neglected. Extracting the constant denominator, Eq. (6.91) becomes

$$\Delta_d^{Trop} = 10^{-6} \, N_{d,0}^{Trop} \frac{1}{h_d^4} \int\limits_{h=0}^{h=h_d} (h_d - h)^4 \, dh \tag{6.92}$$

for an observation site on the earth's surface (i.e., $h = 0$) and gives after integration

$$\Delta_d^{Trop} = 10^{-6} \, N_{d,0}^{Trop} \frac{1}{h_d^4} \left[-\frac{1}{5}(h_d - h)^5 \Big|_{h=0}^{h=h_d} \right] \,. \tag{6.93}$$

The evaluation of the expression between the brackets gives $h_d^5/5$ so that

$$\Delta_d^{Trop} = \frac{10^{-6}}{5} \, N_{d,0}^{Trop} \, h_d \tag{6.94}$$

is the dry portion of the tropospheric path delay at the zenith.

The wet portion is much more difficult to model because of the strong variations of the water vapor with respect to time and space. Nevertheless, due to lack of an appropriate alternative, the Hopfield model assumes the same functional model for both the wet and dry components. Thus,

$$N_w^{Trop}(h) = N_{w,0}^{Trop} \left[\frac{h_w - h}{h_w} \right]^4 \tag{6.95}$$

where the mean value

$$h_w = 11\,000 \text{ m} \tag{6.96}$$

is used. Sometimes other values such as $h_w = 12\,000$ m have been proposed, cf. Fell (1980). Unique values for h_d and h_w cannot be given because of their dependence on location and temperature. Kaniuth (1986) investigated a local situation with radio sonde data over 4.5 years and calculated for the region of the observation site $h_d = 41.6$ km and $h_w = 11.5$ km. The effective troposphere heights are given as 40 km $\leq h_d \leq 45$ km and 10 km $\leq h_w \leq 13$ km.

The integration of (6.95) is completely analogous to (6.91) and thus results in

$$\Delta_w^{Trop} = \frac{10^{-6}}{5} N_{w,0}^{Trop} h_w . \tag{6.97}$$

The total tropospheric path delay at zenith thus is

$$\Delta^{Trop} = \frac{10^{-6}}{5} \left[N_{d,0}^{Trop} h_d + N_{w,0}^{Trop} h_w \right] \tag{6.98}$$

with the dimension meters. The model in its present form does not account for an arbitrary zenith distance of the signal. Considering the line of sight, an obliquity factor must be applied which in its simplest form is the projection from the zenith onto the line of sight given by $1/\cos z$, cf. also (6.59). Frequently, the transition of the zenith delay with $z = 0$ to a delay with arbitrary zenith distance z is denoted as the application of a mapping function, cf. for example Lanyi (1984), Janes et al. (1989), Kaniuth et al. (1989).

A slight variation of the Hopfield model contains an arbitrary elevation angle E (expressed in degrees) at the observing site. Seeber (1989), p. 53 presents the formulas where for the dry component $\sin(E^2 + 6.25)^{-1/2}$ and for the wet component $\sin(E^2 + 2.25)^{-1/2}$ are used as mapping functions:

$$\Delta^{Trop}(E) = \frac{10^{-6}}{5} \left[\frac{N_{d,0}^{Trop} h_d}{\sqrt{\sin(E^2 + 6.25)}} + \frac{N_{w,0}^{Trop} h_w}{\sqrt{\sin(E^2 + 2.25)}} \right] . \tag{6.99}$$

In more compact form, Eq. (6.99) can be represented as

$$\Delta^{Trop}(E) = \Delta_d^{Trop}(E) + \Delta_w^{Trop}(E) \qquad (6.100)$$

where

$$\Delta_d^{Trop}(E) = \frac{10^{-6}}{5} \frac{N_{d,0}^{Trop} h_d}{\sqrt{\sin(E^2 + 6.25)}}$$

$$\Delta_w^{Trop}(E) = \frac{10^{-6}}{5} \frac{N_{w,0}^{Trop} h_w}{\sqrt{\sin(E^2 + 2.25)}} \qquad (6.101)$$

or, by substituting (6.87), (6.90) and (6.88), (6.96), respectively,

$$\Delta_d^{Trop}(E) = \frac{10^{-6}}{5} \frac{77.64 \frac{p}{T}}{\sqrt{\sin(E^2 + 6.25)}} \left[40\,136 + 148.72\,(T - 273.16)\right]$$

$$\Delta_w^{Trop}(E) = \frac{10^{-6}}{5} \frac{-12.96\,T + 3.718 \cdot 10^5}{\sqrt{\sin(E^2 + 2.25)}} \frac{e}{T^2}\, 11\,000$$

$$(6.102)$$

results. Measuring p, T, e at the observation location and calculating the elevation angle E, the total tropospheric path delay is obtained in meters by (6.100) after evaluating (6.102).

Modified Hopfield models. The empirical function (6.89) is now rewritten by introducing lengths of position vectors instead of heights. Denoting the earth's radius by R_E, the corresponding lengths are $r_d = R_E + h_d$ and $r = R_E + h$, cf. Fig. 6.4. The dry refractivity in the form

$$N_d^{Trop}(r) = N_{d,0}^{Trop} \left[\frac{r_d - r}{r_d - R_E}\right]^4 \qquad (6.103)$$

is thus equivalent to (6.89). Applying Eq. (6.84) and introducing a mapping function, gives

$$\Delta_d^{Trop}(z) = 10^{-6} \int_{r=R_E}^{r=r_d} N_d^{Trop}(r) \frac{1}{\cos z(r)}\, dr \qquad (6.104)$$

for the dry path delay. Note that the zenith distance $z(r)$ is variable. Denoting the zenith distance at the observation site by z_0, the sine-law

$$\sin z(r) = \frac{R_E}{r} \sin z_0 \qquad (6.105)$$

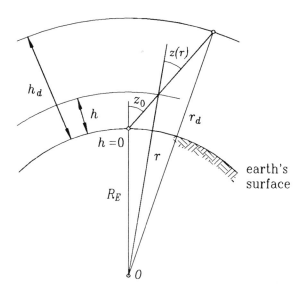

Fig. 6.4. Geometry for the tropospheric path delay

can be applied, cf. Fig. 6.4. From Eq. (6.105) follows

$$\cos z(r) = \sqrt{1 - \frac{R_E^2}{r^2} \sin^2 z_0} \tag{6.106}$$

which is equivalent to

$$\cos z(r) = \frac{1}{r} \sqrt{r^2 - R_E^2 \sin^2 z_0} . \tag{6.107}$$

Substituting (6.107) and (6.103) into (6.104) yields

$$\Delta_d^{Trop}(z) = \frac{10^{-6} N_{d,0}^{Trop}}{(r_d - R_E)^4} \int_{r=R_E}^{r=r_d} \frac{r(r_d - r)^4}{\sqrt{r^2 - R_E^2 \sin^2 z_0}} \, dr \tag{6.108}$$

where the terms being constant with respect to the integration variable r have been extracted from the integral. Assuming the same model for the wet portion, the corresponding formula is given by

$$\Delta_w^{Trop}(z) = \frac{10^{-6} N_{w,0}^{Trop}}{(r_w - R_E)^4} \int_{r=R_E}^{r=r_w} \frac{r(r_w - r)^4}{\sqrt{r^2 - R_E^2 \sin^2 z_0}} \, dr . \tag{6.109}$$

Instead of the zenith distance z the elevation angle $E = 90° - z$ could also be used. Many modified Hopfield models have been derived, depending solely on the method to solve the integral. Among them, Janes et al. (1989) mention for instance the models of Yionoulis (1970), Goad and Goodman (1974), Black (1978), Black and Eisner (1984). Here, one model is presented basing on a series expansion of the integrand. Details can be found for example in Goad and Goodman (1974). The resulting formulas can be found e.g. in Remondi (1984) where a subscript i is introduced which reflects either the dry component (replace i by d) or the wet component (replace i by w). With

$$r_i = \sqrt{(R_E + h_i)^2 - (R_E \cos E)^2} - R_E \sin E \qquad (6.110)$$

the tropospheric delay in meters is

$$\Delta_i^{Trop}(E) = 10^{-12} N_{i,0}^{Trop} \left[\sum_{k=1}^{9} \frac{\alpha_{k,i}}{k} r_i^k \right] \qquad (6.111)$$

where

$$\alpha_{1,i} = 1 \qquad\qquad \alpha_{6,i} = 4a_i b_i (a_i^2 + 3b_i)$$

$$\alpha_{2,i} = 4a_i \qquad\qquad \alpha_{7,i} = b_i^2 (6a_i^2 + 4b_i)$$

$$\alpha_{3,i} = 6a_i^2 + 4b_i \qquad\qquad \alpha_{8,i} = 4a_i b_i^3 \qquad (6.112)$$

$$\alpha_{4,i} = 4a_i (a_i^2 + 3b_i) \qquad\qquad \alpha_{9,i} = b_i^4$$

$$\alpha_{5,i} = a_i^4 + 12a_i^2 b_i + 6b_i^2$$

and

$$a_i = -\frac{\sin E}{h_i}$$

$$\qquad\qquad\qquad\qquad\qquad\qquad (6.113)$$

$$b_i = -\frac{\cos^2 E}{2h_i R_E} .$$

Substituting $i = d$, then the dry part results where in (6.111) for $N_{d,0}^{Trop}$ Eq. (6.87) and for h_d Eq. (6.90) must be introduced. Analogously, Eqs. (6.88) and (6.96) must be used for $N_{w,0}^{Trop}$ and for h_w.

Saastamoinen model. The refractivity can alternatively but equivalently be deduced from gas laws, the interrelationship is demonstrated e.g. in Janes et al. (1989). The Saastamoinen model is based on this approach where again some approximations have been employed. Here, any theoretical derivation is omitted. Saastamoinen (1973) models the tropospheric delay, expressed in meters,

$$\Delta^{Trop} = \frac{0.002277}{\cos z} \left[p + \left(\frac{1255}{T} + 0.05 \right) e - \tan^2 z \right] \qquad (6.114)$$

as a function of z, p, T and e. As before, z denotes the zenith distance of the satellite, p the atmospheric pressure in mbar, T the temperature in Kelvin, and e the partial pressure of water vapor in mbar. Saastamoinen has also refined this model by adding two correction terms, one being dependent on the height of the observing site and the other on the height and on the zenith distance. Bauersima (1983) gives the refined formula as

$$\Delta^{Trop} = \frac{0.002277}{\cos z} \left[p + \left(\frac{1255}{T} + 0.05 \right) e - B \tan^2 z \right] + \delta R \quad (6.115)$$

where the correction terms B and δR can be interpolated from Tables 6.2 and 6.3.

Tropospheric problems. There are many other tropospheric models which are similar to the models given here, e.g. Lanyi (1984), Chao (1972), Marini and Murray (1973), Elgered et al. (1985), Davis et al. (1985), Rahnemoon (1988). Although the list is not complete, the question arises why there are so many different approaches. One reason is the difficulty in modeling the water vapor. The simple use of surface measurements cannot give the utmost accuracy so that water vapor radiometers have to be developed. These instruments measure the sky brightness temperature by radiometric microwave observations along the signal path enabling the calculation of the wet path delay, cf. Elgered et al. (1985). The hardware components of a water vapor radiometer are described, for example, in Reichert (1986). Accurate water vapor radiometers are expensive and experience problems at low elevation

Table 6.2. Correction term B for the refined Saastamoinen model

Height [km]	B [mbar]
0.0	1.156
0.5	1.079
1.0	1.006
1.5	0.938
2.0	0.874
2.5	0.813
3.0	0.757
4.0	0.654
5.0	0.563

Table 6.3. Correction term δR in meters for refined Saastamoinen model

Zenith distance	Station height above sea level [km]							
	0	0.5	1.0	1.5	2.0	3.0	4.0	5.0
60°00′	0.003	0.003	0.002	0.002	0.002	0.002	0.001	0.001
66°00′	0.006	0.006	0.005	0.005	0.004	0.003	0.003	0.002
70°00′	0.012	0.011	0.010	0.009	0.008	0.006	0.005	0.004
73°00′	0.020	0.018	0.017	0.015	0.013	0.011	0.009	0.007
75°00′	0.031	0.028	0.025	0.023	0.021	0.017	0.014	0.011
76°00′	0.039	0.035	0.032	0.029	0.026	0.021	0.017	0.014
77°00′	0.050	0.045	0.041	0.037	0.033	0.027	0.022	0.018
78°00′	0.065	0.059	0.054	0.049	0.044	0.036	0.030	0.024
78°30′	0.075	0.068	0.062	0.056	0.051	0.042	0.034	0.028
79°00′	0.087	0.079	0.072	0.065	0.059	0.049	0.040	0.033
79°30′	0.102	0.093	0.085	0.077	0.070	0.058	0.047	0.039
79°45′	0.111	0.101	0.092	0.083	0.076	0.063	0.052	0.043
80°00′	0.121	0.110	0.100	0.091	0.083	0.068	0.056	0.047

angles, see Lanyi (1984), since the tropospheric path delay at zenith is amplified by the mapping function.

The difficulty in modeling the tropospheric effect will require continuation of research and development for some years. Due to the opinion of Lanyi (1984), the best solution is to combine surface and radio sonde meteorological data, water vapor radiometer measurements and statistics. This is a major task and an appropriate model has not yet been found.

6.4 Relativistic effects

6.4.1 Special relativity

Lorentz transformation. Consider two four-dimensional systems $S(x, y, z, t)$ and $S'(x', y', z', t')$ where the union of space coordinates x, y, z and the time coordinate t is characterized by space-time coordinates. The system S is at rest and, relatively to S, the system S' is uniformly translated with velocity v. For simplicity, it is assumed that both systems coincide at an initial epoch $t = 0$ and that the translation takes place along the x-axis.

The transformation of the space-time coordinates is given by the Lorentz transformation

$$x' = \frac{x - v\,t}{\sqrt{1 - \frac{v^2}{c^2}}} \qquad\qquad x = \frac{x' + v \cdot t'}{\sqrt{1 - \frac{v^2}{c^2}}}$$

$$y' = y \qquad\qquad y = y'$$

$$z' = z \qquad\qquad z = z' \tag{6.116}$$

$$t' = \frac{t - \frac{v}{c^2}\,x}{\sqrt{1 - \frac{v^2}{c^2}}} \qquad\qquad t = \frac{t' + \frac{v}{c^2}\,x'}{\sqrt{1 - \frac{v^2}{c^2}}}$$

where c is the speed of light. An elegant and simple derivation of these formulas can be found in Joos (1956), pp. 217–218. By using Eqs. (6.116), the relation

$$x^2 + y^2 + z^2 - c\,t^2 = x'^2 + y'^2 + z'^2 - c\,t'^2 \tag{6.117}$$

may be verified. This means that the norm of a vector in space-time coordinates is invariant with respect to the choice of its reference system. Note that in case of $c = \infty$ the Lorentz transformation converts to the Galilei transformation

$$x' = x - v\,t$$

$$y' = y$$

$$z' = z \tag{6.118}$$

$$t' = t$$

which is fundamental in classical mechanics.

The theory of special relativity is by definition restricted to inertial systems. The application of the Lorentz transformation reveals some features of that theory.

Time dilation. Consider two time events t_1 and t_2 at the same location x of the system at rest. Due to the Lorentz transformation (6.116), the corresponding events in the moving system take the form

$$t'_1 = \frac{t_1 - \frac{v}{c^2}\,x}{\sqrt{1 - \frac{v^2}{c^2}}} \qquad\qquad t'_2 = \frac{t_2 - \frac{v}{c^2}\,x}{\sqrt{1 - \frac{v^2}{c^2}}}\,. \tag{6.119}$$

Denoting the time interval in the moving system by $\Delta t' = t'_2 - t'_1$ and in the resting system by $\Delta t = t_2 - t_1$, the difference of the two expressions in (6.119) yields the time dilation

$$\Delta t' = \frac{\Delta t}{\sqrt{1 - \frac{v^2}{c^2}}} \tag{6.120}$$

which means that the time interval Δt in S for an observer moving with S' is lengthened to $\Delta t'$. If the time intervals are monitored by clocks, the moving clocks run slower than resting clocks. For the inverse situation the same holds so that an intervall $\Delta t'$ in S' is lengthened to Δt for a resting observer.

Lorentz contraction. The derivation of the Lorentz contraction is analogous to the time dilation. Considering now two locations x_1 and x_2 in the resting system S at the same epoch t, then the corresponding locations in the movig system S' are given by the Lorentz transformation. Using the abbreviations $\Delta x = x_2 - x_1$ and $\Delta x' = x'_2 - x'_1$, then it infers from above that $\Delta x'$ is lengthened to Δx for a resting observer. Expressing it in another way, for a resting observer the dimension of a moving body seems to be contracted:

$$\Delta x' = \Delta x \sqrt{1 - \frac{v^2}{c^2}} \, . \tag{6.121}$$

Second-order Doppler effect. Since frequency is inversely proportional to time, one can deduce immediately from the considerations on time dilation that the frequency f' of a moving emitter would be reduced to f when received by an resting observer. This is the second-order Doppler effect given by the formula

$$f' = \frac{f}{\sqrt{1 - \frac{v^2}{c^2}}} \, . \tag{6.122}$$

Mass relation. Special relativity also affects masses. Denoting the masses in the two reference frames S and S' by m and m', then

$$m' = \frac{m}{\sqrt{1 - \frac{v^2}{c^2}}} \tag{6.123}$$

is the corresponding mass relation, cf. for example Heckmann (1985).

Each of the formulas (6.120) to (6.123) comprises the same square root which may be expanded into binomial series:

$$\frac{1}{\sqrt{1 - \frac{v^2}{c^2}}} = 1 + \frac{1}{2} \left(\frac{v}{c} \right)^2 \cdots$$

$$\sqrt{1 - \frac{v^2}{c^2}} = 1 - \frac{1}{2} \left(\frac{v}{c} \right)^2 \cdots \, . \tag{6.124}$$

Substituting these expansions into Eqs. (6.120) to (6.123), each related to an observer at rest, then

$$\frac{\Delta t' - \Delta t}{\Delta t} = \frac{\Delta x' - \Delta x}{\Delta x} = -\frac{f' - f}{f} = \frac{m' - m}{m} = -\frac{1}{2}\left(\frac{v}{c}\right)^2 \quad (6.125)$$

accounts for the mentioned effects of the special relativity in one formula.

6.4.2 General relativity

The theory of general relativity includes accelerated reference systems too, where the gravitational field plays the key role. Formulas analogous to (6.125) may be derived when one replaces the kinetic energy $\frac{1}{2}v^2$ in special relativity by the potential energy ΔU. Thus,

$$\frac{\Delta t' - \Delta t}{\Delta t} = \frac{\Delta x' - \Delta x}{\Delta x} = -\frac{f' - f}{f} = \frac{m' - m}{m} = -\frac{\Delta U}{c^2} \quad (6.126)$$

represents the relations in general relativity, cf. Heckmann (1985), where ΔU is the difference of the gravitational potential in the two reference frames under consideration.

6.4.3 Relevant relativistic effects for GPS

The reference frame (relatively) at rest is located in the center of the earth and an accelerated reference frame is attached to each GPS satellite. Therefore, the theory of special and general relativity must be taken into account. Relativistic effects are relevant for the satellite orbit, the satellite signal propagation, and both the satellite and receiver clock. An overview of all these effects is given for example in Zhu and Groten (1988), the relativistic effects on rotating and gravitating clocks is also treated in Grafarend and Schwarze (1991). With respect to general relativity, Ashby (1987) shows that only the gravitational field of the earth must be considered. Sun and moon and consequently all other masses in the solar system are negligible.

Relativity affecting the satellite orbit. The gravitational field of the earth causes also relativistic perturbations in the satellite orbits. An approximate formula for the disturbing acceleration is given by Eq. (4.46). For more details the reader is referred to Zhu and Groten (1988).

Relativity affecting the satellite signal. The gravitational field gives rise to a space-time curvature of the satellite signal. Therefore, a propagation correction must be applied to get the Euclidean range for instance. The range correction modeled by Holdridge (1967) takes the form

$$\delta^{rel} = \frac{2\mu}{c^2} \ln \frac{\varrho^j + \varrho_i + \varrho_i^j}{\varrho^j + \varrho_i - \varrho_i^j} \tag{6.127}$$

where μ is the earth's gravitational constant. The geocentric distances of satellite j and observing site i are denoted ϱ^j and ϱ_i, and ϱ_i^j is the distance between the satellite and the observing site. In order to estimate the maximum effect for a point on the earth's surface take the mean radius $R_E = 6\,370\,\text{km}$ and a mean altitude of $h = 20\,200\,\text{km}$ for the satellites. The maximum distance ϱ_i^j results from the Pythagorean theorem and is about $25\,800\,\text{km}$. Substituting these values, the maximum range error $\delta^{rel} = 18.7\,\text{mm}$ results from (6.127). Note that this maximum value only applies to point positioning. In relative positioning the effect is much smaller and amounts to $0.001\,\text{ppm}$, cf. Zhu and Groten (1988).

Relativity affecting the satellite clock. The fundamental frequency f_0 of the satellite clock is $10.23\,\text{MHz}$. All the signals are based on this frequency which is influenced by the motion of the satellite and by the difference of the gravitational field at the satellite and the observing site. The corresponding effects of special and general relativity are small and may be linearly superposed. Thus,

$$\delta^{rel} \equiv \frac{f_0' - f_0}{f_0} = \frac{1}{2}\left(\frac{v}{c}\right)^2 + \frac{\Delta U}{c^2}$$
$$\text{special} + \text{general} \tag{6.128}$$
$$\text{relativity}$$

is the effect on the frequency of the satellite clock where Eqs. (6.125) and (6.126) have been used. To get a numerical value, circular orbits and a spherical earth with the observing site on its surface are assumed. Backing on these simplifications, (6.128) takes the form

$$\delta^{rel} \equiv \frac{f_0' - f_0}{f_0} = \frac{1}{2}\left(\frac{v}{c}\right)^2 + \frac{\mu}{c^2}\left[\frac{1}{R_E + h} - \frac{1}{R_E}\right] \tag{6.129}$$

with v being the satellite's mean velocity. Substituting numerical values yields

$$\frac{f_0' - f_0}{f_0} = 4.464 \cdot 10^{-10} \tag{6.130}$$

which, despite the simplifications, is sufficiently accurate. Ashby (1987) for instance takes into account the J_2-term for the potential and the centrifugal forces and gets the only slightly different result $4.465 \cdot 10^{-10}$. Recall that f_0' is the emitted frequency and f_0 is the frequency received at the observation site. Thus, it can be seen that the satellite transmitted nominal frequency would be increased by $df = 4.464 \cdot 10^{-10} \cdot f_0 = 4.55 \cdot 10^{-3}$ Hz. However, it is desired to receive the nominal frequency. This is achieved by an offset df in the satellite clock frequency, so that 10.22999999545 MHz are emitted, cf. Spilker (1980).

Another small periodic effect arises due to the assumption of a circular orbit (which is almost true for the Block II satellites). An adequate correction formula is given by Gibson (1983) as

$$\delta^{rel} = \frac{2}{c} \sqrt{\mu a}\, e \sin E \tag{6.131}$$

where e denotes the eccentricity, a the semimajor axis, and E the eccentric anomaly. This relativistic effect is taken into account by the clock polynomial broadcast via the navigation message, cf. Sect. 4.4.2, where the time dependent eccentric anomaly E is expanded into a Taylor series. Thus, Eq. (6.131) gives slightly more accurate results. However, the decision to use this formula must be correlated with a correction of the clock polynomial coefficients with respect to the built in accounting for the periodic effect. Van Dierendonck et al. (1980) present the necessary formulas. In case of relative positioning, the effect cancels out, cf. Zhu and Groten (1988).

Relativity affecting the receiver clock. A receiver clock located at the earth's surface is rotating with respect to the resting reference frame at the geocenter. The associated linear velocity at the equator is approximately 0.5 km$\cdot$s^{-1} and thus roughly one tenth of the satellite's velocity. Substituting this value into the special relativistic part of Eq. (6.128) yields a relative frequency shift in the order of 10^{-12} which after 3 hours corresponds to a clock error of 10 nanoseconds (1 ns $= 10^{-9}$ s $\doteq 30$ cm). Ashby (1987) presented the correction formula for this Sagnac effect; however, since usually the correction is performed by the receiver software the formula is not explicitly given here.

6.5   Multipath

The effect is well described by its name: a satellite emitted signal arrives at the receiver via more than one path. Multipath is mainly caused by reflecting surfaces near by the receiver, cf. Fig. 6.5, and secondary effects are reflections off the satellite.

Referring to Fig. 6.5, the satellite signal arrives at the receiver on three different paths, one direct and two indirect ones. As a consequence, the

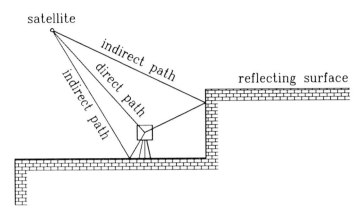

Fig. 6.5. Multipath effect

received signals have relative phase offsets and the phase differences are proportional to the differences of the path lengths, cf. Tranquilla (1986). There is no general model of the multipath effect because of the arbitrarily different geometric situations. However, the influence of the multipath can be estimated by using a combination of $L1$ and $L2$ code and carrier phase measurements, cf. Evans (1986). The principle is based on the fact that the troposphere, clock errors, and relativistic effects influence code and carrier phases by the same amount. This is not true for ionospheric refraction and multipath which are frequency dependent. Taking ionospheric-free code ranges and carrier phases, and forming corresponding differences, all aforementioned effects except for multipath cancel out. The residuals, apart from the noise level, thus reflect the multipath effect.

Among the counter measures against multipath the most effective one is to avoid it. Considering Fig. 6.5, the placing of the antenna directly on the reflecting ground without tripod would eliminate one of the two indirect paths. However, the vertical reflecting surface would still contaminate the results. The recommendation therefore is to avoid as far as possible reflecting surfaces in the neighbourhood of the receivers.

The elimination of multipath signals is also possible by selecting an antenna that takes advantage of the signal polarization. GPS signals are right-handed circularly polarized whereas the reflected signals are left-handed polarized, cf. Scherrer (1985). A reduction of the multipath effect may also be achieved by digital filtering, wideband antennas, cf. Bletzacker (1985), and radio frequency absorbent antenna ground planes, cf. Tallqvist (1985). The absorbent antenna ground plane reduces the interference of satellite signals with low or even negative elevation angles which occur in case of multipath.

Purely from geometry it is clear that signals received from low satellite

elevations are more susceptable to multipath than signals from high elevations. It should be noted that code ranges are more strongly affected by
multipath than carrier phases, cf. Evans (1986). Comparing single epochs,
the multipath effect may amount to 10–20 m for code pseudoranges, cf. Wells
et al. (1987) and Evans (1986). In really bad situations even loss of lock may
occur. However, in case of carrier phases and for relative positioning with
short baselines, the error due to multipath should in general be smaller
than 1 cm for good satellite geometry and a reasonable long observation
interval, cf. Counselman (1981). But even in those cases, as Rocken and
Meertens (1989) show, a simple change of the height of the receiver may
increase the multipath and thus deteriorate the results.

6.6 Antenna phase center offset and variation

The phase center of the antennas is the point to which the radio signal measurement is referred and generally is not identical with the physical antenna
center. The offset depends on the elevation, the azimuth, and the intensity
of the satellite signal and is different for $L1$ and $L2$. Two effects must be
distinguished: the offset and the variation of the antenna phase center. The
precision of an antenna should be based on the antenna phase center variation and not on the offset. A constant offset could easily be determined and
taken into account.

Firstly, the true antenna phase center may be different from the manufacturer indicated center. This antenna offset may simply arise from in-
accurate production series. An investigation to determine this offset was
e.g. performed by Sims (1985) and was based on test measurements by
rotating the antenna in a laboratory environment. Secondly, the antenna
phase varies with respect to the incoming satellite signals. The variation
is systematic and may be investigated by test series. Sims (1985) found
variations amounting to 1–2 cm. However, it is fairly difficult to model the
antenna phase center variation because it is different for each antenna and
also for various types. Geiger (1988) shows the different characteristics of
conical spiral antennas, microstrip antennas, dipole antennas (crossed pair
of horizontal, half-wavelength dipole), and helices. As a consequence, the
direct computation of the antenna effects on the distance measurements with
respect to azimuth and elevation was proposed. Simple functions for an appropriate modeling may also be found by laboratory tests, cf. Schupler and
Clark (1991).

7. Surveying with GPS

7.1 Introduction

This chapter mainly concerns the practical aspects of GPS surveying and addresses planning, performance, and in situ data processing. Overlapping with other chapters is intentional to provide complete information in a single chapter for readers more interested in practical considerations.

7.1.1 Terminology definitions

The increasing interest in GPS is reflected by the numerous papers published today. Unfortunately, a standard terminology is not used at this time; although, several authors, e.g. Wells (1985), have attempted to provide a list of terms used. To avoid confusion in the following sections, some of more important definitions are given here and are used throughout the entire text. Also, some alternative terminology frequently used in scientific papers is listed at the end of this section. However, a thorough list is not presented.

Code range versus carrier phase
Typically, GPS observables are pseudoranges derived from code or carrier phase measurements. Generally speaking, the accuracy of code ranges is at the meter level whereas the accuracy of carrier phases is in the millimeter range. However, the accuracy of code ranges can be improved by smoothing techniques. Unlike the carrier phases, the code ranges are unambiguous. The ambiguities of the phases can be determined by various methods (e.g., initialization procedures).

Real-time versus postprocessing
To qualify as real-time GPS, the position results must be available in the field immediately or while still on station. The results are denoted as "instantaneous" if the observables of a single epoch are used for the position computation and the processing time is negligible. A different and less stringent definition is "real-time" which includes computing results after more than one observation epoch. The original concept of GPS is the feature of instantaneous navigation which is the determination of the position of a moving vehicle (i.e., a ship, a car, an aircraft) by unsmoothed code pseudoranges, cf. for example Yiu et al. (1984).

Postprocessing occurs when data are collected and processed after the fact. Typically, the data are not processed in the field and measurements from different sites are combined.

Point positioning versus relative positioning

The coordinates of a single point are determined for point positioning using a single receiver which measures code ranges to (normally four or more) satellites.

Relative positioning is possible if two receivers are used and (code or carrier phase) measurements, to the same satellites, are simultaneously made at two sites. The accuracy is better than in the case of point positioning as a consequence of processing the data from two sites. Normally, the coordinates of one site are known and the position of the other site is to be determined relatively to the known site (i.e., the vector between the two sites is determined). In general, the receiver placed on the known site is stationary while observing. In a more stringent definition the term "relative" is used in the case of carrier phase observations whereas in the case of code range observations the term "differential" is used, cf. for example Strange (1985); Oswald et al. (1986); Blackwell (1986).

Another definition associates point positioning with navigation and relative positioning with surveying.

Static versus kinematic

Static denotes a stationary observation location while kinematic implies motion. A temporary loss of signal lock in static mode is not as critical as in kinematic mode. The terms "static" and "kinematic" must be considered in the context of point or relative positioning. Typical examples of these modes are given to acquaint the reader with these terms.

Static point positioning is useful if points are needed with moderate accuracy, say 5 m to 10 m, after a fairly short observation time. Numerical examples are given e.g. in Wells and Lachapelle (1981), Lachapelle et al. (1982), Ashjaee (1986).

Kinematic point positioning can be used to determine a vehicle's trajectory in space and time with an accuracy of 10 m to 100 m, cf. Cannon et al. (1986). Some examples are vehicle navigation, airborne gravimetry, and gravity vector determination using a land vehicle, cf. Schwarz (1987).

Static relative positioning by carrier phases, at present, is the most frequently used method by surveyors and is also called *static surveying*. The principle is based on determining the vector between two stationary receivers. This vector is often called "baseline" or simply "line" because of its similarity to triangulation baselines. According to this terminology the process is called single or multipoint baseline determination. Obviously, the multipoint solution concerns more than two sites. In static surveying 1 ppm to 0.1 ppm accuracies are achievable which is equivalent to millimeter accuracy for baselines up to some kilometers.

Kinematic relative positioning involves one stationary and one moving receiver. The two receivers perform the observations simultaneously. The possible applications are basically the same as for kinematic point positioning but the accuracies are higher. The accuracy in differential positioning (i.e., by code ranges) is at the meter level, and in relative positioning (i.e., by carrier phases) centimeter accuracy is achievable, cf. Cannon et al. (1986).

Additional definitions

Semikinematic, cf. Cannon and Schwarz (1989), or *stop and go*, cf. Minkel (1989), is a combination of the static and kinematic relative positioning. This mode is characterized by alternatively stopping and moving one receiver with main interest in the stopped positions. The most important feature of this method is the increase in accuracy when several measurement epochs at the stop locations are accumulated and averaged. This technique is often referred to as simply kinematic survey.

Pseudokinematic or *intermittent static* is a method developed by Remondi (1988). The points of interest are revisited after about one hour to facilitate ambiguity resolution and to obtain a better accuracy (subcentimeter level). This is mainly accomplished by the changed satellite configuration. There is no requirement for maintaining signal lock between the reoccupation of points. The receiver may even be turned off while moving.

Rapid static techniques use code and carrier phase combinations for a fast initialization (i.e., ambiguity resolution) in static mode. This technique requires measurements of both code and carrier phase on both frequencies. Observations of 5 to 10 minutes can produce 1 ppm survey accuracy.

On-the-fly (OTF) techniques carry out the initialization in kinematic mode rather than in static mode. This code aided technique will permit positioning of moving vehicles with decimeter ore even centimeter accuracy once the ambiguities are resolved.

Alternative definitions

This brief section mentions terminologies used sometimes by various authors. Actually, in some cases these notations are clear, but in other cases also very confusing.

Instead of *point positioning* the term *single point positioning*, cf. for example Lachapelle et al. (1982) or Prilepin (1989), or the term *absolute point positioning*, cf. for example Gouldman et al. (1986) or Hartl et al. (1985), is used. Here, the term *absolute* reflects the opposite of *relative*.

Attention should be paid to the difference between the terms *kinematic* and *dynamic*. A very intuitive example given in Schwarz et al. (1987) points up the difference:

"Modeling the movement of a vehicle in three-dimensional space requires either the knowledge of the forces causing the motion or the measurement of the vehicle motion in a given three-dimensional coordinate system. The first type of modeling will be called dynamic, the second kinematic."

The modeling of the orbit for GPS satellites is a dynamic procedure. Hein et al. (1988) consider this as the only case of dynamic modeling with respect to the analysis of GPS observations. As soon as the positions of the satellites are assumed to be known and given, positioning of a moving vehicle can be considered as kinematic procedure.

7.1.2   Observation technique

The selection of the observation technique in a GPS survey depends upon the project's particular requirements; especially the desired accuracy plays a dominant role.

When using a single receiver, only point positioning with code pseudoranges makes sense. One should keep in mind that unlimited access is only provided to the C/A-code by the Standard Positioning Service (SPS) and that the accuracy can be deliberately degraded by turning on selective availability (SA). The Precise Positioning Service (PPS) offers access to both codes but is limited to authorized users when the system is declared operational and anti-spoofing (A-S) or encryption of the P-code will be invoked.

When two or more receivers are used, one known site serves as monitor station and higher accuracies are achievable at the second site. In the differential mode, code pseudoranges are simultaneously observed to (normally) four or more satellites. The monitor station calculates the actual corrections to the observed code pseudoranges. These corrections are then transmitted by various communication links to the unknown sites leading to an improvement of their independently computed positions. With P-code receivers providing accuracies of the code ranges at the meter level, differential positioning with submeter accuracy is achievable. Hence, surveys of wetlands and other areas requiring less than third-order accuracy can be economically accomplished using the differential code ranging technique. The advantage of a code pseudorange is its immunity from cycle slips (i.e., changes of the phase ambiguities) and to some extent from site obstructions.

Consequently, code range measurements in wooded area are less affected by the tree branches than carrier phase measurements.

At present, geodetic accuracies are only obtained in the relative positioning mode with observed carrier phases. Processing a baseline vector requires the phases be simultaneously observed at both baseline endpoints. Hence, relative positioning originally was only possible by postprocessing data. Recently, successful attempts of real-time data transfer on short baselines have been made which enable real-time computation of a baseline vector, cf. for example Hofmann-Wellenhof et al. (1990).

The static surveying method is the most commonly used since the only basic requirement is a relatively unobstructed view of the sky for the occupied points. Static surveys generally, require 60- to 120-minute observation periods. However, this method also includes the shorter (e.g., 10-minute) observation wide-laning technique or rapid static surveying based on a fast ambiguity resolution approach, cf. Frei and Beutler (1990). For long baselines (e.g., ≥ 50 km) the detection of cycle slips and the resolution of ambiguities may become difficult. In such cases it may be useful to run additional receivers in the vicinity of the two baseline sites. The ambiguities are fixed using the short baselines and can then be used to fix the ambiguities for the long baseline. The procedure is also called "boot-strapping", cf. Stangl et al. (1991). Typical uses of static surveying include: state, county, and local control surveys; photo-control surveys, boundary surveys; and deformation surveys.

Kinematic surveys are the most productive in that the greatest number of points can be determined in the least time. Whereas the static GPS method requires that the satellites move in the sky, the kinematic method does not require this motion. Because of this fact, the proposed interoperation of GPS with geostationary satellites will be useful for kinematic surveys, cf. Sect. 13.3.3. Kinematic surveys require considerable reconnaissance since not only the occupied sites but also the route taken to travel between sites must be free of obstructions. The kinematic technique requires that lock be maintained on four or more satellites throughout the entire survey. This means, in practical terms, that the roving receiver cannot travel under a tree or too close to a utility pole. The kinematic method is best suited for wide open areas (such as cleared construction sites) where there are few obstructions. Areas such as suburban subdivisions can also be surveyed by the kinematic technique if there are not too many large trees overhanging the roads. The kinematic technique can be used to position a receiver that is mounted on an all-terrain vehicle that travels across a given area in a series of cross section lines. The three-dimensional coordinates of this vehicle mounted receiver can be determined to high accuracy (few centimeters) so that an accurate topographic map of the area can be prepared.

The technique most nearly like the static method is the pseudokinematic survey. These surveys require less occupation time but one must occupy the "point-pair" two times. A typical scenario would involve occupying a pair of points for five minutes, moving to other points, and finally returning to the first point pair about one hour after the initial occupation for a second five minute occupation. The advantage of the (total) 10-minute occupation versus the 60-minute occupation is somewhat offset by the time lost in traveling to the points a second time. The best use of the pseudokinematic method is when the points to be occupied are along a road where the observers can move quickly between setups. The main advantage of the pseudokinematic method is that for a certain observation time more sites can be occupied than with the conventional static surveying. Compared to the kinematic method, a loss of lock may occur and the number of satellites does not play that essential role. The main weakness is the necessity of the site revisiting. This restricts the method to local applications. The main competitor is therefore the rapid static positioning approach where each site has to be occupied only once. Typical pseudokinematic applications include: photo-control, lower-order control surveys, and mining surveys.

In practice, it is best to use a mixture of the three methods. For example, static and pseudokinematic methods can be used to establish a broad framework of control and to set points on either side of obstructions such as bridges. Kinematic surveys can then be employed to determine the coordinates of the major portion of points, using the static points as control and check points. A thorough reconnaissance is required for these mixed surveys.

In order to serve as a datum reference for subsequent surveys and to enable the transformation of GPS results into national datums, two types of GPS control networks are in use; namely, passive control networks and active control systems. The passive concept is to tie GPS control networks to existing triangulation monuments and vertical benchmarks. The disadvantage is that many sites have to be occupied and maintained. However, this system is appropriate where dense national triangulation networks exist and when the control network serves other purposes such as geodynamical investigations. Such a dense passive control network is being established e.g. in Austria as reported by Stangl et al. (1991).

The objectives of an active control system are the computation and (near) real-time dissemination of differential corrections for single-receiver users, and the computation of precise ephemerides for postmission processing. The collection and dissemination of the data is performed by the use of high-speed terrestrial, and satellite-based communication links. More details on this subject are provided in Sect. 12.2.2.

7.1.3   Field equipment

The field equipment includes receiver units and auxiliary devices such as meteorological sensors, tribrachs, tripods, bipods, and other ancillary equipment. Here, geodetic receivers which perform precise baseline vector measurements are mainly considered. The selection of an appropriate receiver depends on the special requirements in a project. Therefore, only some general considerations will be given in this section.

For short baselines up to about 30 km, single frequency receivers can be used because the influence of the ionospheric refraction (mostly) cancels by differencing the phase measurements between the baseline sites. Baseline distances must be reduced in periods of high sunspot activity. This activity cycle has a repetition rate of about 11 years with a maximum in early 1991. Dual frequency receivers compensate (and virtually eliminate) ionospheric refraction by the ionospheric-free combination of the two carrier phases, cf. Sect. 6.3.2.

An important aspect of receiver design is the data sampling rate. A fast rate produces a large volume of data and requires a significant amount of storage. This fast rate is necessary for kinematic applications, and in static techniques it facilitates cycle slip detection and repair. Also, the use of receivers with more than the minimum four channels is appropriate since additional tracked satellites deliver redundant information.

Receivers with P-code capability are essential for the most precise positioning applications. Using P-code survey receivers that measure both code ranges and carrier phases, the baseline vectors can be determined more accurately and rapidly than with other types of receivers. One reason is that P-code receivers guarantee a better code range resolution and that they are less affected by multipath and imaging. Another reason is that the P-code enables the reconstruction of both carriers by the code correlation technique. The disadvantage of P-code receivers is that when anti-spoofing is turned on, the P-code will be encrypted and unavailable to civilian users. In this event, the $L1$ carrier is obtained by C/A-code correlation and the reconstruction of the $L2$ carrier is performed by a codeless technique such as squaring.

Another important feature for receiver selection (particularly for kinematic surveys) is its capability of bandwidth selection in the tracking loops. The bandwidth should be wide enough to prevent loss of signal, but narrow enough to provide a high signal-to-noise ratio. Therefore, receivers that are able to adapt the bandwidth depending on the dynamics will provide optimum results.

Different receiver types involved in a survey may cause problems due to incompatibility (e.g., different number of channels, different signal processing techniques). The receiver time tagging of observations is of critical importance. The interpolation of measurements to a common reference epoch proposed by Gurtner et al. (1989) may prove difficult because of the dithering of the time signal when SA is turned on. One method to ensure common time tagging is to use external oscillators as proposed by Landau (1990).

The phase center of the antenna should be stable and repeatable which seems to be best realized in microstrip antennas. When different antenna types are used, the phase center of each type must be calibrated. It should be noted that the phase center depends on frequency and is therefore different for $L1$ and $L2$ carrier observations, cf. Prescott et al. (1989). Most of the GPS receiver manufacturers offer coaxial antenna cables with different lengths up to 60 m where 10 m is standard. A long cable provides more versatility for site access; however, manufacturers' recommendations on cable size (type) should be followed to avoid signal loss. Large diameter low loss cable is available for tracking sites where the cable length exceeds 60 m.

Today, a number of weather-proof, lightweight (4–5 kg), small (6 000–7 000 cm^3), and low power consuming (less than 10 W) receivers which are capable of tracking "all-in-view" satellites (6–12 channels) are readily available. New technologies such as MMIC (Monolithic Microwave Integrated Circuits) or VHSIC (Very High Speed Integrated Circuit) will further miniaturize the receivers as pointed out by Cannon (1990).

All GPS equipment being marketed today gives excellent results so the decision of which equipment to use is usually made based upon ease of use and monetary considerations. The cost of GPS receivers is decreasing rapidly and their capability is constantly increasing. The TI-4100 produced in 1984 was priced at \$ 170 000. Today, a simple C/A-code receiver costs about \$ 20 000 and this price will still decrease to less than \$ 10 000 in the near future. When comparing equipment costs, the user should determine if the processing software cost is included. An estimation on the initial capital and start-up costs for three GPS receivers and related items such as software and computers is less than \$ 100 000, cf. Henstridge (1991).

The final factor that should be considered is the primary use of the equipment. For example, if one plans to use the equipment for kinematic surveys or in heavily treed areas, a receiver should be chosen that has a separate antenna that can be mounted on a prism pole or survey mast. Equipment with built-in antennas have optional antennas that can be purchased. Another important feature if kinematic surveys on foot are planned, is the weight of the receiver and its power requirements. A heavy, very power consuming receiver is not the type of instrument to carry for extended periods.

The number of surveying receiver manufacturers is constantly increasing and their products are continuously being improved. To get the actual information on a particular product, the reader is referred to contact the manufacturer directly. The addresses are provided in regularly compiled lists such as by Arradonda-Perry (1992).

7.2 Planning a GPS survey

7.2.1 General remarks

Designing a GPS network gives rise to several practical questions which are just as important as the theoretical aspects of GPS as stated by Bevis (1991). The issues of equipment, observation technique, and organization are all important. In this section the last item will be addressed.

GPS surveying differs essentially from classical surveying because it is weather independent and there is no need for intervisibility between the sites. Because of these differences, GPS surveys require different planning, execution, and processing techniques. At present, only a few organizations such as U.S. Federal Geodetic Control Committee (FGCC) (1988) have published standards and specifications for GPS surveys.

Planning a GPS survey may or may not be appropriate for a particular application or project. Before deciding whether or not to expend resources on planning, one must decide the ultimate purpose of the survey and the accuracy of the desired results (coordinates). The FGCC classifies surveys by designating orders for the various proportional accuracies desired. Generally, surveys of the highest order (A and B) are reserved for special purposes such as state and county control networks and scientific studies. The remaining first-, second-, and third-order surveys are the classifications normally specified for mapping, surveying, and engineering projects. There is no question that a substantial planning effort is required to successfully complete a high-order survey; however, first-order and lower-order surveys may not require extensive planning except in heavily treed or obstructed areas.

The optimum planning of a GPS survey has to consider several parameters such as site or satellite configurations, the number and the type of receivers to be used, and economic aspects. Contrary to the design of triangulation or trilateration networks which involves considerable effort to maintain the geometric strength, for GPS networks geometry and line length are not so critical. The planning phase should also include some data processing considerations, for instance whether the available software allows for the computation of single baseline vectors or for multipoint solutions.

For large projects with many sites and many receivers, planning a GPS survey could be aided by the use of computer programs to save time and resources. The Geodetic Survey of Canada for instance has developed an appropriate software package, cf. Klees (1990). This program plans vehicle routes, the selection of satellites, and the delineation of the network design.

7.2.2   Presurvey planning

Point selection. The first step in planning a GPS survey is to obtain the largest scale map of the area upon which the desired points can be plotted. Topographic maps from 1:25 000 to 1:100 000 scale are excellent for this purpose, also county road maps are quite useful. All desired survey points are plotted on the map along with the known control points. In most instances it is not worth using control points that are not included in the national system. Using the coordinates of points of unknown accuracy can create many problems so that it is better to choose a national network point even if it is at a greater distance from the project site than other control monuments.

In planning a GPS survey, there are only two basic considerations in choosing a point: (1) its location in an area of good sky visibility, and (2) its proximity to a road. The first requirement is of primary importance; however, the proximity to a road is only a convenience that increases production. Following is a list of the desirable GPS site characteristics.

1. Clear view of the sky above 20° elevation.

2. Easily accessible (preferably by vehicle).

3. Mark not likely to be disturbed.

4. Clear site for visible azimuth mark.

5. Space for parked vehicle.

6. On publicly owned land.

The complete tie of a GPS network to the national datum requires the occupation of three or more control points. Many areas now have supernets which provide convenient reference monuments for GPS surveys. If supernet points are available, one may not even have to visit them prior to beginning field work since the network points have been selected for GPS occupation. A reasonable horizontal tie would consist of measurements to national control points on either side of the project area. Ties to the vertical datum normally require more planning. Three benchmarks at the corners of the project area would provide the minimum acceptable vertical datum reference.

Observation window. The second step of presurvey planning is to determine the optimum daily observation period and to decide how it should be subdivided into sessions (i.e., periods where two or more receivers simultaneously track the same satellites).

The optimum "window" of satellite availability is the period when a maximum of satellites can be observed simultaneously. The length of the window is a function of the location and, at present, is shortened by the limited satellite constellation. The optimum window is found by inspecting azimuth-elevation charts which are produced by the software of GPS equipment manufacturers.

The calculation of azimuth and elevation is based on the projection of the unit vector $\Delta \underline{\varrho}$, pointing from the observing site to the instantaneous satellite position, onto the orthogonal axes of the local coordinate frame. The unit vector $\Delta \underline{\varrho}$ is defined by

$$\Delta \underline{\varrho} = \frac{\underline{\varrho}^S - \underline{\varrho}_R}{\|\underline{\varrho}^S - \underline{\varrho}_R\|} \tag{7.1}$$

with $\underline{\varrho}^S$ being the geocentric position vector of the satellite and $\underline{\varrho}_R$ being the geocentric position vector of the observing site which is needed only approximately (i.e., coordinates from a map are sufficient). The vector $\underline{\varrho}^S$ can be calculated by procedures described in Chap. 4 and the vector $\underline{\varrho}_R$ corresponds to the vector $\underline{X}$ defined by Eq. (3.6).

With φ and λ being the ellipsoidal latitude and longitude of the observing site, the axes $\underline{i}$, $\underline{j}$, $\underline{k}$ of the local coordinate frame are given by

$$\underline{i} = \begin{bmatrix} -\sin \varphi \cos \lambda \\ -\sin \varphi \sin \lambda \\ \cos \varphi \end{bmatrix} = \frac{\partial \underline{k}}{\partial \varphi}$$

$$\underline{j} = \begin{bmatrix} -\sin \lambda \\ \cos \lambda \\ 0 \end{bmatrix} = \frac{1}{\cos \varphi} \frac{\partial \underline{k}}{\partial \lambda} \tag{7.2}$$

$$\underline{k} = \begin{bmatrix} \cos \varphi \cos \lambda \\ \cos \varphi \sin \lambda \\ \sin \varphi \end{bmatrix} .$$

Hence, adequate equations for the zenith distance z and the azimuth a, reckoned from north towards east positively, follow from the inner products,

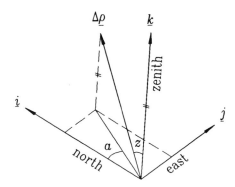

Fig. 7.1. Satellite's zenith distance z and azimuth a

cf. Fig. 7.1,

$$\Delta\underline{\varrho} \cdot \underline{i} = \sin z \cos a$$

$$\Delta\underline{\varrho} \cdot \underline{j} = \sin z \sin a \qquad\qquad (7.3)$$

$$\Delta\underline{\varrho} \cdot \underline{k} = \cos z \,.$$

Table 7.1 is part of a typical tabular listing of satellite locations. The leftmost column shows the Universal Time Coordinated (UTC) and each subsequent column shows the vertical (elevation) angle $E = 90° - z$ and azimuth a for several satellites.

A graphical representation of the complete elevation-azimuth list, based on an elevation cut off angle of 20°, is given in Fig. 7.2. This figure indicates the satellites in view (shown on the ordinate of the coordinate system and indicated as space vehicle SV with the corresponding PRN number) for a certain epoch. No matter what observation technique is used or the precision desired, only periods with four or more satellites in view is considered to be viable satellite coverage. Therefore, in the example given the time spans 04:00–10:00 UTC and 14:00–22:00 UTC are not recommended for observation.

An improved representation of visibility is given by the "sky plots". These polar or orthogonal plots show the satellite path in function of elevation angle and azimuth, cf. Fig. 7.3. Such sky plots are often supplemented by time tags and by the image of the local horizon. The latter may be obtained from field reconnaissance (e.g., fish-eye photos) or can be generated by the computer from a digital terrain model.

Apart from visibility, the tracked satellites should be geometrically well distributed with (ideally) one in each of the four quadrants. This is only a guideline however, and not a firm requirement. In static surveys, poor sat-

Table 7.1. Part of elevation E azimuth a list for Palm Beach,
Florida, on April 15, 1991

SV	2		11		14		15		18		19	
UTC	E	a	E	a	E	a	E	a	E	a	E	a
19:00			24	160	29	243			35	318	10	295
19:20			33	156	24	234			43	314	15	301
19:40			43	151	18	226			51	306	21	308
20:00			53	144	13	219			57	294	27	314
20:20			62	132	8	212	1	174	60	276	33	320
20:40	0	258	69	112	3	205	5	168	60	258	40	326
21:00	5	265	71	79			12	163	56	238	48	331
21:20	9	273	67	51			18	157	50	225	56	336
21:40	13	280	59	36			25	151	42	215	65	342
22:00	18	288	50	30			33	145	34	208	74	349

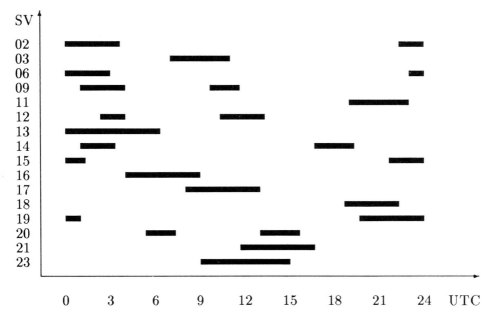

Fig. 7.2. Satellite visibility for Palm Beach, Florida, on April 15, 1991 (cut
off elevation: 20 degree)

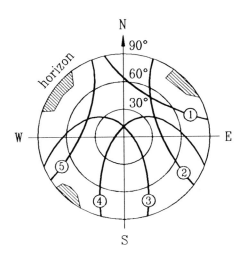

Fig. 7.3. Polar sky plot

ellite geometry and even lack of the fourth satellite can often be offset by observing for a longer period of time. Movement of the satellites with respect to each other improves the geometry and thus the solution. Manufacturers' recommendations should be followed especially regarding the use of three satellites, since it is critical to ensure that the observation time tags are correct. The surest method of selecting periods of adequate satellite coverage is to make test observations over known (or determined) baseline vectors for the full period being considered. This large data set can then be divided into smaller subsets for processing during periods being considered for observations. This method allows one to test the acceptability of both the specific period of coverage and the chosen length of observation time. A measure for satellite geometry is the GDOP (Geometric Dilution of Precision) factor. Normally, GDOPs under six are considered good and those above six are considered as being too high.The GDOPs reflect only the instantaneous geometry related to a single point. Therefore, factors for baseline vectors, accumulated over the time span of a session, are more appropriate precision indicators, cf. Merminod et al. (1990). It has been proposed that quality factors can be computed by the receiver itself, and the selection of the optimum configuration could be automated. Appropriate formulas for GDOP calculation are provided in Sect. 9.5 where also details on the decomposition of GDOP into several components such as PDOP (Position Dilution of Precision) are given.

Another aspect for the selection of the window concerns the ionospheric refraction. Observations during night hours may be appropriate because the ionospheric effect is usually quieter during this time. Normally, however, daylight hours are preferred for organizational reasons.

Sessions. The specific time period chosen for an observation is called a session. Considering for example Fig. 7.2, the period from 01:00 to 02:00 UTC would be suitable for observation since five satellites are available. If this were the first observation of the day it would be designated session "a". The second session (e.g., 02:30 to 03:30 UTC) would be designated session "b". Some manufacturers use number designation but numbers have the disadvantage of requiring two digits for sessions in excess of 10. Normally, the session designator begins with "a" again each day, and days are expressed as the consecutive calendar day (1 to 365 or, in leap years, 366). For example, session 105c means the third session of day 105.

A good time to begin the first session of static surveys is when four or more satellites are above the 15 to 20 degree elevation angle, and the last observation of this session should, generally, end when the fourth satellite drops below 15 to 20 degrees. This is only a general rule since three-satellite time prior to rise of the fourth satellite and after the fourth satellite has set is useful. There are five factors that determine the length of a particular observation. These are:

1. The relative geometry of the satellites and the change in the geometry.

2. The number of satellites (effects geometry).

3. The degree of ionospheric disturbance (for single frequency receivers) worse for higher latitudes and during daylight.

4. The length of the baseline.

5. The amount of obstructions at the sites.

In general, the more satellites that are available, the better the geometry, and the shorter the length of observation required. The length of a session may also be reduced in the case of shorter baselines. For example, sessions with lines 1–2 km in length could be as short as 45 minutes with five satellites ($L1$ receiver). Longer lines between control points might, on the other hand, require 90 minutes of data to achieve good results. For single frequency receiver, Table 7.2 can be used as a general guide to plan the length of observation when four or more satellites are available and the ionospheric conditions are normal.

What is the reason for needing these relatively long sessions? Chap. 6 discusses the observables and it is seen that at each observation epoch the

Table 7.2. Session length in function of baseline length

Baseline [km]	Session [min]
0.1 – 1.0	10 – 30
1.1 – 5.0	30 – 60
5.1 – 10.0	60 – 90
10.1 – 30.0	90 – 120

carrier phase is measured to millimeter accuracy or even better. In effect, a single observation would be sufficient to provide the precision required for a geodetic survey. The difficulty is that one can measure the decimeters, centimeters, and millimeters precisely; however, the observation must last long enough to determine the meters by resolving the integer number of cycles. On a short (less than 1 km) baseline, the integer cycles can often be resolved in 5–10 minutes using $L1$ only phase. With $L1/L2$ P-code receivers using the wide-laning technique, long (15 km) lines can be accurately measured with as little as ten minutes data.

The best method of determining optimum observation times for large projects is to make longer than normal observations on the first day to obtain typical data sets. For example, observations lasting 90 minutes for short (1–5 km) lines and 120 minutes for longer (5–20 km) lines would be made. These data sets when processed would yield excellent results. The observations could then be reprocessed using portions of the data set to determine the point where good results can no longer be obtained. For example, consecutive 30-minute data sets could be processed and compared with the full data set to determine if the shorter observation times were sufficient to achieve good results.

As discussed previously, a session should be long enough to guarantee the required accuracy; but, one should also consider that longer sessions cost more. In any case, the time between the sessions should be long enough to transport the equipment to another site and allow for accurate setup. Older receivers may also require oscillator warm up.

In order to reference the single sessions to a common datum, at least one site of the network must be occupied during the entire project (pivot point concept), or subsequent sessions must contain at least one reoccupied site (leap frogging concept). The reoccupation of more than one site improves the precision and reliability of the network. When planning sessions for kinematic surveys, there are two factors to consider. Normally, times are selected when five or more satellites are above 20 degrees and when the satellites have a GDOP of less than six. For most locations, the GDOP condition is satisfied when five or more satellites are available. Problems occur when one of the satellites is obstructed during the survey and the remaining four satellites have a high GDOP. This problem is solved by keeping the roving receiver stationary until the fifth satellite is reacquired.

Nonplanned surveys. Before proceeding with descriptions of the planning steps, a brief discussion of nonplanned surveys will be given. Not all surveys require extensive planning. Some surveyors are now using GPS as they would use other survey equipment and are not necessarily planning a geodetic "campaign". A good example of a survey requiring a minimum of planning is certain types of photo-control surveys. Since many types of GPS surveys require that points be placed in locations where there may be substantial obstructions, the major planning effort is to layout a scheme and select sites that are relatively free from obstructions. Some types of photo-control surveys do not present the typical obstruction planning problem.

If the area being mapped is a suburban or residential area with clearings well-scattered throughout the area, one can perform the GPS (and photo) point selection during the actual survey. In this type of area, the only reconnaissance or presurvey activity required is to locate both horizontal and vertical control used to reference or tie the survey to the national datum.

Assuming that a photo-control project is performed in a fairly obstruction free area and that acceptable horizontal and vertical control points are nearby, the field crew can be sent to the area with the approximate desired photo-control sites plotted on a county road map or some other relatively small-scale map. A good plan is to begin the survey in a portion of the project area where the sites are close together so the project manager can quickly coordinate the start of the survey. During the first session, the observers setup on their assigned points while the project manager selects and "monuments" the next set of points. The monumentation may be an iron pin driven into the ground or it may be a nail driven into the pavement. Following selection of the second set of points and completion of the first session, the GPS observers are instructed to proceed to the second session, and the project manager selects the third set of points. Upon conclusion of the local scheme, the ties to existing control can be made to complete the survey.

A second example of nonplanned GPS surveys would be the establishment of control at a construction site. Many times, survey crews are sent to construction sites with plans, coordinate lists, and other supporting data; and the crew chief makes decisions and plans the field work on-site. The same procedure could be followed when using GPS equipment. The crew chief would design the network on-site and establish points in areas where they were needed. Immediately following the observations, the data can be transferred from the GPS receivers to a laptop computer (for example in the survey vehicle). Modern laptop computers are able to process GPS data in a matter of minutes per line so that the data from several sessions (occupations) could be completed in less than one hour. The crew chief would then have accurate (first-order) coordinates of control on-site which could be used to layout desired construction stakes.

7.2.3   Field reconnaissance

After the GPS points have been plotted on a map and descriptions of how to reach the existing control have been obtained, one is ready to perform a field reconnaissance. This is also a good time to assign each point a unique identifier. The most obvious method is to consecutively assign each point a number. Points can have more descriptive (full name) identifiers as well, but a simple consecutive number facilitates future reference to each point. The reconnaissance surveyor visits each site to check its suitability based on the factors listed in Sect. 7.2.2.

First of all, static GPS surveys need an unobstructed view of the sky above an elevation of $15° - 20°$ and a nonreflective environment. This is a critical requirement for kinematic applications where the path of the roving antenna should be selected carefully in advance. Easy access of the site is desired to save time between the sessions. This may be less important when cross-country vehicles are used or by observing eccentric sites. The latter may often become necessary in case of forests or urban areas. Sites that have many obstructions require additional consideration. At these sites the reconnaissance surveyor should prepare a polar plot showing the vertical angle and azimuth to obstructions over 20 degrees. This plot is then overlayed the polar sky plot of the satellites, cf. Fig. 7.3. The obstruction problem is solved in two ways. The first is to place the antenna on top of a survey mast so that the desired 20 degree visibility angle is obtained. The Geodetic Survey of Sweden for instance has devised 30 meter (guyed) survey masts that are quickly erected and plumbed over the mark by two offset theodolites. Several manufacturers produce prism poles that extend to 10 meters which also can be used for this purpose. The second technique for overcoming the problem of obstructions is to choose a time when a sufficient number of satellites is electronically visible at the site. The example in Fig. 7.3 shows shaded portions that depict areas obstructed by trees, buildings, hills, etc. It is seen that these obstructions do not effect the observations at this one site. The data used to process a baseline vector consists of the common observations between two points; so the same obstruction check must be made for both ends of a line. Blockage of a satellite at one end of a line effectively eliminates that satellite from the solution; so care must be used in performing the analysis. The manual method of making this analysis is to produce satellite sky plots for every hour of the useable satellite span and then visually compare the site obstruction plots with the various one-hour plots. Another more automated method is to use the software produced by the various manufacturers to perform this analysis on the computer.

Field reconnaissance is a must prior to conducting a kinematic survey. Each site must be checked for sky visibility and the route taken to travel between points also must have good sky visibility. Since the kinematic method requires that lock be continuously maintained on four or more satellites, good sky visibility practically means obstruction free (above 20 degrees vertical angle) situations. When obstructions on the route of travel (such as bridges) occur, static points can be placed on either side of the obstruction so that the roving receiver can be reinitialized. The path that the surveyor is to follow between points should be clearly marked on a large-scale map to make sure that unwanted cycle slips do not occur.

Reconnaissance for pseudokinematic surveys is not as important because this technique only requires the revisited sites being unobstructed. In the case of differential (navigation) surveys where code ranges are measured, reconnaissance is not critical at all because the receivers can be simply turned on and measurements made when desired. This mode would be used more as a precise navigation system than a survey system.

Apart from obstructions, it is important to consider the multipath problem. Multipath (more fully discussed in Sect. 6.5) is the effect of unwanted reflected satellite signals that are received by the antenna. This problem is most severe when the antenna is placed near a chain link fence or other metal structure. The satellite signals are reflected by the metal structure and corrupt the direct signals causing phase errors. In the case of chain link fences, the antenna can be elevated above the fence to eliminate the problem. When the point is close to a metal building, the only practical solution is to move the point to another location. Multipath does not appear to be a problem for points located in the median of highways where large trucks pass by at high speed. The multipath caused when a truck is near the antenna is of too brief duration to cause significant problems. However, one should avoid having a large metal transport truck parked next to the antenna.

When the site meets all requirements, the point can be marked for the marksetters to set the monument. The reconnaissance surveyor should plot the location of the chosen site on the largest scale map of the area and prepare a preliminary "to reach" description that describes how to reach the point from a known prominent location (e.g., local post office). This description will save hours in wasted time in the future since it will allow the marksetters and observers to quickly find the point. In cases where a different crew sets the actual monument, this mark setting crew completes the "to reach" description adding the final specific location of the mark and ties to nearby prominent objects. Surveys performed for inclusion in national networks require that a "to reach" description be prepared, and in some cases that a sketch of the site be made. The final site documentation should also include photos of the station surroundings, name and address of the owner, preliminary coordinates, power supply, etc.

7.2.4   Monumentation

The monumentation normally set eventually will become unnecessary when active control networks have been established. For the present, monumentation is still specified for projects where the sites are planned to be reoccupied (e.g., geodynamical studies), cf. Avdis et al. (1990). Note that monuments set other than in solid rock or by deep concrete pillars are not adequate for high precision surveys.

Monumentation is a general term to describe any object used to mark a point. Land surveyors and others commonly use sections of steel reinforcing bar upon which a cap is crimped to monument a point. Each surveyor must decide which particular type of mark is appropriate for the project. A steel rod may be appropriate to mark a photo-panel point; whereas, a massive concrete monument would be more appropriate for a county geodetic survey. The main consideration is the mark should be easily found, at least for the duration of the survey.

7.2.5   Organizational design

The planning phase for static surveys ends with the layout of an organizational design for the project. First, each field crew is allocated a number of personnel and appropriate vehicle and survey equipment. Each crew is then assigned sites to occupy during specific sessions. Field crews should be fully acquainted with the area and should be able to move quickly between points. Today, the operation of GPS receivers no longer requires highly qualified survey personnel. Still, malfunctions are better solved by trained crews.

The minimum number n of sessions in a network with s sites and using r receivers is given by

$$n = \frac{s - o}{r - o} \tag{7.4}$$

where o denotes the number of overlapping sites between the sessions. Equation (7.4) makes only sense for $r > o$ and $o \geq 1$. In the case of a real number, n must be rounded to the next higher integer.

Another approach for the design implies that each network site should be occupied m times. In this case the minimum number of sessions equals

$$n = \frac{m\,s}{r} \tag{7.5}$$

where n again must be rounded to the next higher integer.

The number s_r of redundant occupied sites with respect to the minimum overlapping $o = 1$ is given by

$$s_r = n\,r - [s + (n - 1)]. \tag{7.6}$$

Consider a squared network consisting of nine sites $(P1, \ldots, P9)$ equally spaced and assume that three receivers (A, B, C) are available. If the "leap frogging" method with one-point overlapping between the sessions is the observation technique selected, Eq. (7.4) gives $n = 4$ as minimum number of sessions. The first four sessions in Table 7.3 represent one possible solution for the corresponding organizational design where the sites $P4, P5$, and $P6$ are reoccupied. When each point should be occupied twice then $m = 2$ and Eq. (7.5) gives $n = 6$ which is reflected by two more sessions in Table 7.3. The chosen design has the property that all short baseline vectors between adjacent sites are observed which provides a homogeneous accuracy in the network. For the elimination of receiver or antenna biases, an interchange of the equipment between the sessions is recommended.

The organizational design also depends upon the type of the network (i.e., the distribution of the sites), cf. Snay (1986) or Unguendoli (1990). There are two basic types of GPS networks: (1) radial, and (2) closed geometric figures.

Table 7.3. Principles of organizational design

Receiver	Session	a	b	c	d	e	f
A		P2	P4	P4	P6	P7	P1
B		P5	P5	P1	P3	P8	P2
C		P8	P6	P7	P9	P9	P3

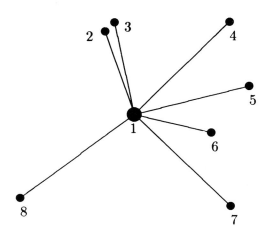

Fig. 7.4. Radial survey

Radial surveys. Radial (or cartwheel) surveys are performed by placing one receiver at a fixed site, and measuring lines from this fixed site to receivers placed at other locations. A typical radial survey configuration is shown in Fig. 7.4. There is no geometric consideration for planning this type of survey except that points in close proximity should be connected by direct observation. Consider in Fig. 7.4 the points 2 and 3 located 10 km from fixed point 1 and 100 m distant from one another. If they were surveyed at different times, the error between the two points could be considerable. A relative baseline error of 10 ppm, for example, produces a 0.1 m error in the points 2 and 3. Therefore, the expected error between the two points would be 0.14 m (as obtained from $\sqrt{0.1^2 + 0.1^2}$) which corresponds to a relative error of 1:714. This large relative error would not be tolerated by most surveyors since it would give a bad impression on the accuracy of the overall survey.

In general, kinematic surveys are radial surveys, and many pseudokinematic surveys are performed in the radial mode. Each point established by the radial method is a "no check" position since there is only one determination of the coordinates and there is no geometric check on the position. An appropriate use of radial surveys might be to establish photo-control since the photogrammetrist is able to make an independent check on the coordinates using analytical bridging. Other uses would be to provide positions for wells or geological features where precise coordinates may generally not be needed.

Network survey. GPS surveys performed by static (and pseudokinematic) methods where accuracy is a primary consideration, require that observations be performed in a systematic manner and that closed geometric figures be formed to provide closed loops. Figure 7.5 shows a typical scheme consisting of 18 points. The preferred observation scheme is to occupy adjacent points consecutively and traverse around the figure. For example, the scheme shown in Fig. 7.5 would be approached in the following manner if three receivers were used. Receiver *A* would be placed on point 1, receiver *B* on point 2, and receiver *C* on point 3 for the first session. Data would be collected from the three receivers for the chosen session length of time. The receivers would then be turned off and receiver *A* would be moved to point 11, and receiver *B* to point 10, while receiver *C* remains on point 3 to overlap between the first two sessions. Following the period during which the receivers were being moved, the three receivers would be turned on for the second session and data collection would be continued. An alternative to this approach is to move all three receivers so that receiver *A* would occupy point 3, receiver *B* point 11, and receiver *C* point 10 for the second session. Either plan works equally well. However, the second approach would eliminate eventual receiver biases, since the overlapping point 3 is occupied by different receivers in consecutive sessions. The leapfrog traversing technique is the preferred method because it provides the required loop checks and gives the maximum productivity. In general, the goal in devising a measuring scheme is to directly connect as many adjacent points as possible.

When national datum coordinates and elevations are desired for points in a scheme, ties to existing control must be made. In Fig. 7.5 horizontal control points are marked by triangles and denoted HCA, HCB, HCC. Proper connection to control points requires that direct measurements be made between the existing control. The purpose of this measurement is to both verify the accuracy of the existing control and to determine scale, shift, and rotation between the control and the new GPS network. After the horizontal control has been occupied, measurements are made between the control and the nearest network point. For example, in Fig. 7.5 points HCA, HCB, and HCC would be occupied during one session and the tie between HCA and point 1 would be made during another session (using three receivers).

Ties to vertical control points are performed somewhat differently than ties to horizontal points. There is no geometric reason for measuring between existing benchmarks (vertical control) since the error in the geoidal height masks any possible GPS error. Vertical ties are normally made from the benchmark to the closest GPS point. In Fig. 7.5, the benchmarks are shown as BMA, BMB, BMC, and BMD. In this figure, BMD was selected to be point 1, thus eliminating the need of a tie for this point. Existing monuments

should be chosen as GPS points, whenever possible, to save placing a mark
in the ground. In the figure, BMA and BMC were assumed to be close to
points 6 and 13, respectively, so that a conventional level (loop) tie was made
between benchmark and GPS point. Benchmark BMB was assumed to be
far enough away from point 18 so that a GPS measurement was used to
connect the two points. In the case of this benchmark, FGCC specifications
call for two determinations of the vector (18–BMB).

Figure 7.5 is an idealized scheme; however, the basic principles apply for
all control network schemes. In summary, the following are the major points
of network design:

1. The network should consist of closed loops or other geometric figures.

2. Ties should be made to at least three (2 supernet) horizontal control
   points which should also be directly occupied.

3. Ties should be made to at least four vertical control points (benchmarks) by the most direct means.

In addition to these general specifications, other measurements may be
prescribed by various agencies, such as making a certain percentage of repeat
measurements and setups.

In the case of one-point overlapping per session and using three receivers,
a network consisting of 22 sites (i.e., 18 GPS points, 3 horizontal control
points, 1 benchmark) would require 11 sessions where one baseline is measured twice, cf. Eqs. (7.4) and (7.6). With 15 sessions, nine loop closures can
be formed. One possible occupation schedule for determining the coordinates
of the 22 points is listed in Table 7.4.

Under the assumption of 60-minute sessions and 30 minutes for site
change, the fifteen sessions would require two days with the present constellation under ideal conditions. A more realistic schedule would be to take
three days to make the sessions longer than one hour.

7.3 Surveying procedure

The actual execution of GPS surveys is greatly facilitated by good planning.
A well-planned survey normally progresses quite smoothly. This section
will discuss the various aspects of conducting a GPS survey and provide
information to make the effort more productive.

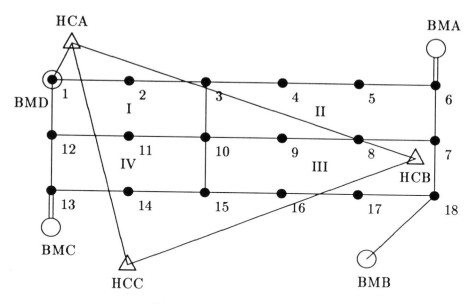

Fig. 7.5. Static network design

Table 7.4. Organizational design for the configuration in Fig. 7.5

Session	Receiver		
	A	B	C
a	HCA	HCB	HCC
b	HCA	1	2
c	3	10	2
d	11	10	12
e	1	13	12
f	HCC	13	14
g	15	10	14
h	3	4	5
i	6	7	5
j	9	7	8
k	9	10	15
l	16	17	15
m	BMB	17	18
n	BMB	HCB	18
o	8	7	18

7.3.1   Preobservation

Antenna setup. To avoid the multipath or imaging effects, it is recommended that any vehicle be parked as far away as possible from the antenna. Parking the survey vehicle the full 10 m (antenna cable length) away from the point should be adequate.

The antennas are mounted on pillars, tripods (with tribrach), or on a range (or prism) pole equipped with steadying legs. This range pole with legs is called a bipod and its use can materially speed up the conduct of a survey. When performing kinematic or pseudokinematic surveys, the bipod is essential so that the antenna is maintained at a fixed height, and to speed the setup time. The use of a bipod for static surveys is not critical; however, its use reduces the possibility of an undetected blunder in measuring the antenna height (since the height of the rod is fixed).

A rather minor problem using GPS should not be overlooked: the calibration of tribrachs. A survey will only be as accurate as the surveyor's ability to center the antenna over the survey point. The best way to avoid this problem is to use two-piece collimating tribrachs that can be rotated to check the centering. A more cost effective approach is to hang a plumb bob from the tribrach at each setup to check the optical centering device.

The antenna phase center offset discussed in Sect. 6.6 is due to the fact that the geometric center of the antenna is not at the electronic center. This problem is virtually eliminated by pointing all antennas in the same direction. In the surveying mode, differences between points are measured so that any systematic offset will be canceled by uniform antenna orientation.

The measurement of the antenna's phase center above the reference point is an important aspect which is often overlooked. Experience has shown that the mismeasured antenna height is the single most vexing problem in conducting GPS surveys and causes the greatest number of errors. The best way to avoid this problem (when using tripods) is to measure the antenna height twice, at the beginning and after completion of the survey. Some manufacturers are providing special rods to facilitate antenna height measurements. When using tripods, the setup at "swing" or repeat points should be broken during the move and the tripod reset over the point at a different height.

Receiver calibration. In general, GPS receivers are considered to be self-calibrating and users do not normally perform equipment calibration. One simple test that can be performed, however, is a zero baseline measurement. This measurement is made by connecting two or more receivers to one antenna. Care should be taken in doing this to use a special device that blocks the voltage from all but one receiver being fed to the antenna. Also, a signal splitter must be used to divide the incoming signal to the multiple receivers.

A normal session (e.g., 60 minutes) is observed and the baseline is computed in the normal manner. Since a single antenna is used, the baseline components should all be zero. This measurement essentially checks the functioning of the receiver circuits and electronics and is a convenient method of trouble shooting receiver problems independent from antenna biases. The zero baseline test is also one way to satisfy specifications which call for equipment calibration.

Initialization. In static surveying, the initialization of some receivers requires the preprogramming or the on-site input of parameters. Some of these are: the selection of the sampling rate, the bandwidth, the minimum number of satellites to track, the start and stop time for the session, the cut off elevation angle, assignment of a data file. Most of the modern receivers have several channels and track all satellites in view. A preselection is thus only necessary to disregard a satellite. The almanac data are generally stored or are gathered after first lock on, otherwise the data for at least one satellite must be keyed in. For (now obsolete) codeless receivers the synchronization of the receiver clocks must be accomplished prior to starting the survey, and an observation program must be copied to the receiver computer.

In the kinematic mode, the phase ambiguities which convert the phase observable into a range are determined during initialization prior to the survey. This is accomplished by three different techniques. First, one can start from a short known baseline which allows ambiguity resolution after a few minutes observation. Another method is to perform a static survey to determine the vector between the fixed point and the unknown starting point for the kinematic survey. The third method is to perform an antenna swap between the fixed point and starting point. Following Hofmann-Wellenhof and Remondi (1988), the antenna swap is accomplished by placing receiver A at the fixed point and receiver B on the starting point. After a few minutes of observation, both receivers are left running and receiver A is moved to the starting point and receiver B is moved to the fixed point. During this move, both receivers must continuously track a minimum of four (preferably more) satellites. The antenna swap is completed by moving receiver A back to the fixed point and receiver B back to the starting point. This antenna swap will determine the vector between the two points to millimeter accuracy for short lines. Logistically, if the fixed receiver antenna were at a remote location such as the roof of the surveyor's office, the survey starting point at a given site would be located by a static survey. On the other hand, if the fixed point were at the project site, the starting point could be placed near (i.e., 10 m) the fixed site and surveyed by the antenna swap method.

7.3.2   Observation

Communications between survey crews are desirable and generally increase efficiency. Even though communication can be critical, today, the major portion of geodetic surveys, where lines average 20 km in length, are performed without communication between observers. In these cases when an observer is late reaching the point, the lines to that point may have to be resurveyed. The use of portable telephones may eliminate this problem since the schedule could be revised on the spot if all parties could communicate. Land (as opposed to geodetic) GPS surveys normally measure lines averaging 5 km so that normal FM transceivers (hand held radios) can be used to communicate between observers. The need for communications is most critical when conducting pseudokinematic surveys because it is important that all receivers collect data during the same time span (except a fixed full time running reference receiver is invoked).

Most static observations can be performed in an automated mode so that an operator is not required, cf. for example Pesec and Stangl (1990). However, it is good practice to perform data checks during the session and any irregularity should be noted in the field log. Also, an operator is often needed to keep the receiver from harm or loss. When conducting geodetic surveys (campaigns) of highest accuracy, meteorological data (wet and dry temperature, air pressure) should be collected during the session at each site as far above the ground as possible. In extreme cases, backscatter radiometers may be used to measure the atmospheric water vapor. The inclusion of observed meteorological data does not improve the results for short baselines as opposed to using standard tropospheric refraction models. The data may possibly be used for future studies.

In kinematic surveying applications, after initialization, the fixed receiver and the roving receiver are placed on the fixed and (now known) starting point for a few minutes of observation. Then the roving receiver proceeds to the points for which coordinates are desired. As long as four or more satellites (with low PDOP) are continuously tracked by both receivers, vectors from the fixed point can be measured to a high degree of accuracy. If (in the case of four satellites) the signal lock is lost or a cycle slip occurs, the initialization must be repeated. This may happen due to shadowing (i.e., obstruction) of the satellite signal by buildings, bridges, trees or other objects. In this case, the roving receiver must return to a point of known position for a reinitialization. In practice, points surveyed should be visited twice so that a check on the determination can be made. Also, several points whose coordinates are known (i.e., from static surveys) should be included in the survey, when possible, to provide additional checks.

The roving receiver normally remains at a point for a few minutes so that observations can be averaged to obtain a more accurate position. A second way to perform kinematic surveys is to alternate the fixed and roving receivers by leapfrogging units. For example consider consecutive points $1, 2, 3, \ldots, n$ along an open highway. Receiver A would be placed at point 1 and receiver B at starting point 2 near point 1. The two receivers would perform an antenna swap to obtain the starting coordinates of point 2. Receiver B would then move to point 3, with receiver A remaining at point 1. Receiver B would then remain fixed at point 3 while receiver A was moved to point 4. In this way, the vectors 2–3, and 3–4 will be measured. The survey could continue in this manner building a geometric traverse of kinematically surveyed points.

Fog or rain does not influence data transmission. However, lightning strikes may destroy the instrument. Therefore, it is recommended to switch off the receivers and to disconnect the antenna during thunderstorms.

7.3.3   Postobservation

At the completion of a session, a check of antenna's position including a remeasurement of its height above the reference is recommended. Older instruments with external recording devices require the labeling of the data storage media.

A final recommendation is to prepare a site occupation sheet that should be completed at the end of the session. This sheet should contain at least the following information:

1. Station Name.

2. Station identifier used for file name.

3. Observer's name.

4. Receiver and antenna serial number.

5. Height of antenna.

6. Start and stop times.

7. Satellites tracked.

8. Navigation position.

9. Problems experienced.

In addition to this minimum amount of information, it is good practice to make a "rubbing" of the top of the survey mark or take a photograph of the mark and to prepare a site sketch of the mark's location (or check existing sketch).

7.3.4   Ties to control monuments

The practice of connecting a GPS network to a horizontal control point is most simple when the control point is in a clear location so that the GPS antenna is placed directly upon the mark. In many cases, however, this is not possible because the control monument is located in an area with many obstructions or where multipath is a problem. In these instances, the point must be occupied eccentrically.

Figure 7.6 shows a typical eccentric occupation. In this figure it is assumed that the control point C is obstructed and that the GPS receiver can be placed on A, a nearby eccentric point. A conventional azimuth mark and second auxiliary point B are also shown. When an azimuth mark exists at the control point, the problem is simplified because the azimuth to the eccentric point can be determined by measuring the angle at the control point between the azimuth mark and the eccentric point. An alternative method would be to measure the astronomic (third-order) azimuth to the eccentric point. A second auxiliary point such as point B should be used as a check on the measured distance from the control point to the eccentric point and all angles and distances should be measured in the triangle (ABC). A second technique can be used when there is no azimuth mark at the control point. Referring to Fig. 7.6, one GPS antenna would be placed on the eccentric mark A and a second receiver placed upon point B. In this case, the dis-

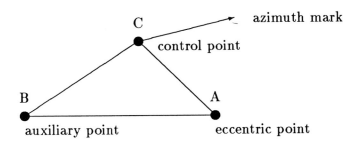

Fig. 7.6. Eccentric occupation

tance between the eccentric and the auxiliary points should exceed the distance from the eccentric point to the control station. From the GPS baseline vector the azimuth of the line $(A–B)$ can be derived and the azimuth $(A–C)$ to the control point can be computed by the angle measured at A. The azimuth from the control point to the eccentric point is determined in one iteration by subtracting the convergence from the reverse azimuth.

Occupying a benchmark eccentrically is accomplished by setting an eccentric point in a clear area near the benchmark and double running levels between the two marks. It should be remembered when occupying horizontal control points eccentrically, where the control point has an accurate elevation, to level between the marks.

7.4 In situ data processing

7.4.1 Data transfer

Most modern equipment stores the GPS observations internally while some of the older instruments write the observations to floppy disk or tape. The first step in processing is to transfer the data from the receiver to a computer hard disk. This transfer is accomplished using software provided by the manufacturer. Observation files for a given session contain the phase and other observables as the main file, in addition broadcast ephemerides and site data consisting of station identifier, antenna height, and possibly navigation position. The main task in transferring files is to make sure the files are named correctly and that the antenna height is correct.

A good procedure to ensure that the file names and antenna heights are correct is to prepare an abstract form, cf. Table 7.5, containing the assignment of sites to sessions, the measured antenna heights, and the start and stop times of the sessions.

Many times an observer will enter the incorrect site identifier. This identifier must be immediately corrected before proceeding with the processing. The controlling document should be the site occupation sheets from which the information in Table 7.5 was abstracted. With this information one knows for instance that point 001 was occupied by receiver B from 02:29 to 03:30 UT. If this point were misidentified, one could look at the start and stop times in the given data file also noting the navigation latitude and longitude. This start and stop time and position contained in each data file pinpoints that particular file to a given time and location which can then be matched with the data in Table 7.5. Using this technique one can identify

Table 7.5. Control abstract

Session	Receiver						Time (UT)	
	A		B		C		Start	Stop
	Site	Height	Site	Height	Site	Height		
a	HCA	1.234	HCB	1.574	HCC	1.342	01:00	02:00
b	HCA	0.987	001	1.782	002	1.543	02:29	03:30
c	003	1.344	010	1.328	002	1.452	04:01	05:00
d	011	1.324	010	1.563	012	1.437	05:31	06:30
e	001	1.564	013	1.453	012	1.455	06:59	08:00

and rename files that have been misnamed.

Most of the batch file processing software automatically extracts the antenna height from the site file data stored in the receiver. After correcting the names of the various files, one should then check and correct all antenna heights. Again a form such as Table 7.5 will be helpful in keeping track of the various antenna heights. Also, a table listing the various occupations and antenna heights is useful for inclusion in the project report.

As soon as all files have been corrected, the observed data should be backed up on at least two sets of storage medium (i.e., floppies). A good practice is to keep one set of original data in a different secure location. Today, many organizations keep a complete set of storage medium in a vault.

7.4.2 Data processing

In the case of remote and extended GPS surveys, a quality control check of the data at least once per day should be made. This control may also include cycle slip detection and repair which, when not performed in the receiver during operation, is normally part of the preprocessing procedure, see Sect. 9.1. The quality control check could also include preliminary computations of baseline vectors in the field before leaving the survey area. The on-site vector processing on a daily basis helps that adequate measurements are being made.

Today, most of the routine processing is performed by batch file processing. All batch files are generated using the three or four digit site identification so that the first task in processing GPS data is to ensure that all sites are properly named. A good practice is to first number the control points and then consecutively assign numbers to adjacent points. This numbering should have been accomplished during the planning and reconnaissance phase to facilitate record keeping.

Processing of static surveys. Modern processing software uses batch processing to compute baseline vectors. Normally, the data for a given day are loaded into a subdirectory on a hard disk. Processing software is normally in another directory and the "path" of the computer has to be set to access the programs. Once the software has been initiated (generally by menu commands), the lines are computed in order, automatically. There are two types of processing software: (1) vector by vector, and (2) multipoint solutions.

The vector by vector or single baseline solution type is presently the most common, and in any case should be used prior to processing with the multipoint software. In some instances one of the points in an observation session will be corrupted and if all points are processed together, the errors from the bad point are distributed among the vectors and the error is masked. The single vector software provides a better check on bad lines or points. The bad point can be more easily isolated by noting that the statistics (i.e., root mean square error, standard error) to lines leading to this one point are worse than the statistics for the other lines. Additionally, the vectors can be summed for the lines in the session, and if the sum around the perimeter is not a small value (e.g., 1 ppm), this indicates that one of the points in the session is bad.

The individual vector processing software performs the following steps:

1. Computes best fit value for point positions from code pseudoranges.

2. Creates undifferenced phase data from receiver carrier phase readings and satellite orbit data. Time tags may also be corrected.

3. Creates differenced phase data and computes their correlations, cf. Sects. 8.2.1 and 8.2.2.

4. Computes estimate of vector using triple-difference processing. This method is insensitive to cycle slips but provides least accurate results.

5. Computes double-difference solution solving for vector and (real) values of phase ambiguities.

6. Estimates integer values of phase ambiguities computed in step 5, and decides whether to continue with fixed ambiguities.

7. Computes fixed bias solution based upon best ambiguity estimates computed in step 6.

8. Computes several other fixed bias solutions using integer values differing slightly (e.g., by 1) from selected values.

9. Computes ratio of statistical fit between chosen fixed solution and the next best solution. This ratio should be at least two to three indicating that the chosen solution is at least two to three times better than next most likely solution.

Processing of kinematic surveys. The basic steps are similar for static and kinematic processing. The data files are downloaded from the receiver to the computer and the file names and antenna heights are checked. The actual processing differs depending on the software used; however, much of the newer software is automated so that hands-on interaction is not needed. The main check for kinematic vectors is to compute positions of the roving receiver and check that similar values are obtained on separate visits to the same point. Also, it is good survey practice to visit points whose coordinates are known during the survey as a further check of the method.

Differential code range processing. Processing code pseudorange data is accomplished using software provided by manufacturers and third party vendors. Most programs are automated so that only the precise position of the fixed station is entered and the coordinates of the second point are computed. The data seems to have a strong central tendency so that averaging several observations rapidly improves the determined position. Positions can be determined for either fixed or moving receivers. Most uses of this technique are radial surveys so that no check of the results is possible.

7.4.3 Trouble shooting and quality control

Single baseline vectors. There are various quality numbers that indicate how well the survey was performed. The first analysis involves inspecting the statistics of individual vectors. The key for determining which vectors are bad is to compare the statistics of a good line with those of a bad line.

The three baselines shown in Table 7.6 are actual data collected in Texas by single frequency receivers. The first and third line were of 30 minutes duration, and the second line lasted one hour. For all three lines, five satellites were in view for nearly the entire period. The sampling rate was 20 seconds.

The results were abstracted from the output or solution files. The first column in Table 7.6 indicates the type of solution: triple-difference (TRP), double-difference with real valued (float) ambiguities (FLT), and double-difference with the ambiguity biases fixed to integers (FIX). The next three columns are the components of the baseline vector (i.e., differences of ECEF coordinates, cf. Chap. 3). The fifth through the seventh columns are the standard errors of the three coordinate differences listed in columns two through four. Column eight (single number) is the ratio between the chosen

Table 7.6. Baseline statistics

1	2	3	4	5	6	7	8	9
	DX	DY	DZ	σ_{DX}	σ_{DY}	σ_{DZ}	Ratio	rms
	[m]	[m]	[m]	[m]	[m]	[m]		[m]
TRP	−303.457	135.317	158.292	1.419	1.011	0.520		0.003
FLT	−303.431	135.314	158.284	0.068	0.062	0.027		0.003
FIX	−303.437	135.327	158.263	0.003	0.006	0.004	30.1	0.004
TRP	2890.453	−594.500	−477.233	0.743	0.366	0.356		0.002
FLT	2890.383	−594.520	−477.221	0.028	0.017	0.012		0.004
FIX	2890.382	−594.526	−477.215	0.002	0.007	0.002	233.1	0.004
TRP	−191.888	−343.451	−546.721	6.686	0.699	1.488		0.004
FLT	−192.221	−343.366	−546.721	0.538	0.069	0.102		0.007
FIX	−192.217	−343.192	−546.689	0.027	0.107	0.051	1.7	0.052

fixed bias solution and the next best solution. The last column is the root mean square (rms) error of fit in meters.

The basic line statistics in Table 7.6 indicate both good baselines and problem baselines. The single most important number in the solution output is the ratio shown in column eight. This number is a strong indicator of both good and bad lines, and should be greater than three, especially for short baselines up to about 5 km. When the ratio is a large number as in the second solution, the line is an accurate measurement. In this case, the difference between the fixed and float solutions is small and the change in the rms between the fixed and float solutions is not dramatic. The second baseline is included in Table 7.6 as an example of a good line with "normal" statistics. The third baseline in Table 7.6 is a bad line which is clearly shown by several indicators. First, note that the ratio (column 8) is 1.7 compared to a similar line with a ratio of 30.1. Both baselines one and three are of similar length and both have observations to satellites for 30 minutes. The standard error of the vector components (columns 5 to 7) are substantially greater for the third line although the number of observations was actually greater for this baseline. The critical numbers in Table 7.6 are the differences between the fixed and float solutions. Note that the difference in the DY-value for the third baseline is 0.174 m. In the computation of this line, the software default was chosen to compute the fixed bias solution even though the ratio did not indicate the fixed solution would be a valid one. This "forced" computation of the fixed bias often yields valid solutions, but particularly on short lines with brief observations, the forced solution yields bad results. Another indicator that the observations do not fit the solution model is the large jump in the rms of fit shown in the last column. The rms of the third baseline jumped from 0.007 to 0.052, a sevenfold increase. Compare this change to the rms of the first line which shows the normal slight increase of

Table 7.7. Float ambiguity solution

First baseline		Third baseline	
Satellites	Ambiguities	Satellites	Ambiguities
12-24	−970431.089	03-16	−184540.781
13-24	−832899.977	17-16	−155781.542
16-24	2235113.884	20-16	29969.388
20-24	61256.060	24-16	12931.837

the three types of solutions.

The statistics of the third line clearly indicate that the fixed bias solution is invalid. This fact is further demonstrated by the inconclusive bias values. The computed ambiguities are compared in Table 7.7. Satellite 24 and satellite 16 are the reference satellites for the double-differences of the first and the third baseline, respectively. It is clearly seen that the ambiguities of the first baseline are obvious by inspection since the values are close to integers. On the other hand, the integer biases of the third line are not obvious. The bias (17-16) could be rounded either up or down and the biases (03-16), and (20-16) are too close to 0.5 to be clearly defined. It is a good practice, when in doubt, to inspect the ambiguities to determine if there is an obvious problem. If the measured line is short and you have at least an hour of adequate observations, one can try reprocessing the line using a different reference satellite, or try eliminating satellites with troublesome bias values. In the case of line three in this example there was simply an insufficient amount of data.

Networks. For networks the best single tool for finding problem lines is to use loop closure software that sums the vector components around a loop to determine misclosures. Most GPS software packages have this type of software.

The first step in checking loops is to prepare a simple sketch showing the measured lines. Depending upon the software, the lines should then be numbered to correspond with the data file used by the software, or in some cases the path through the network can be chosen by referring to certain node points in the network. Bad lines in a loop are found by computing different combinations that both include and exclude the bad line. For example, if the line from point 10 to 11 (Fig. 7.5) were a bad line, it would be discovered by a high misclosure in the loops I and IV. When the larger loop around both small loops is computed (i.e., 1–3–15–13–1), the loop would have an acceptable closure. Therefore, the bad line could be narrowed to either line 10–11, or line 11–12. Normally, one would then inspect the output files for these lines more closely to determine if one or the other was more suspect. At this point, either the suspect line would be remeasured or point 11 would be dropped or downweighted.

The second quality control procedure in a network is to compute a minimally constrained least squares adjustment (often called free adjustment). This adjustment should normally be performed only after bad lines have been eliminated using the loop closure programs. There are numerous (PC compatible) adjustment programs available today, and possibly many more in the future. Chap. 9 gives a more complete description of least squares adjustment and only some practical uses will be discussed here. Each software package contains programs that prepare the necessary input file automatically from the various manufacturers' GPS output files.

Least squares adjustment programs perform three basic tasks: (1) they first shift all vectors so that they are connected in a contiguous network, (2) they add small corrections to each vector component to obtain a "flat" geometric figure closure, and (3) they compute coordinates and elevations of all points. The network design is important in obtaining useful information from an adjustment. A single loop would provide the minimum required information while a network such as the one in Fig. 7.5 would provide a high degree of redundancy needed for network analysis.

In the batch processing of GPS data, the navigation position is often used as the starting coordinate for the vector computation and the solution is linearized around this value. When performing a minimally constrained adjustment, the datum coordinates of one point are normally entered, and the coordinates of all vectors are shifted to agree with the chosen coordinates of the one point. Therefore, a listing of the amounts each point was shifted to be consistent with the fixed point is available and can be printed. It is good practice to scan this list of shifts to determine if the starting coordinates for any of the vectors was in error by more than a reasonable amount. For example, if one of the shifts were 200 meters, you would probably wish to recompute that particular vector using a better starting coordinate (e.g., obtained from preliminary adjustment). Also, these shifts give a good indication of how well the C/A-code pseudorange positions agree with the datum position.

The proper weighting of observations (vectors) is always a problem. Many of the adjustment programs build a weight matrix based upon the vector correlation matrix and standard errors in the GPS output file. The formal standard errors of the vector computation are generally optimistic

by a factor of three to ten times. Therefore, the weight matrix composed
of these optimistic values must be scaled to arrive at a true estimate of
the network errors. The problem is to determine what scale factor to use.
Experience with the various software packages will show how large a scale
factor is appropriate. Also, the adjustment software or the GPS software
documentation should indicate the various scaling factors to use.

Adjustment software that does not use the vector correlation matrix
requires the a priori estimates of the vector errors. The allowable error for
first-order (1:100 000) surveys according to the FGCC specifications at the
two sigma confidence level is 10 mm + 10 ppm of the distance. This error
can be chosen as the a priori value for the adjustment, but first must be
halved to express the value in terms of the one sigma or standard error level.
When this 5 mm + 5 ppm a priori error is used, and the adjustment yields a
standard error of unit weight of one or less, one knows that the work meets
first-order specifications.

Another method for checking the quality of the GPS network is to inspect the residuals resulting from the adjustment. Normally, there are two
residual lists. One lists the actual amounts that vector components have
been corrected to achieve (exact) closure of the network. These residuals
should consist of small values for short lines and larger values for longer
lines. Blunders or large errors will be distributed throughout a general area
and may be difficult to isolate. A better way to isolate large errors is to use
the loop closure software in combination with the adjustment programs.

The second list of residuals provided by most programs comprises the
normalized or standardized residuals. These values are unitless and are
the actual residuals scaled. A value of 1.0 for a normalized residual would
indicate that the residual was as large as expected using the a priori weight
model. A value less than one indicates the residual is less than expected
and a value greater than one indicates a larger than expected value. In an
adjustment a few normalized residuals as high as 2.0 are to be expected.
When more than 5% of the residuals are greater than 2.0 and for any value
over 3.0, further investigations should be performed, and some lines possibly
eliminated.

A typical case where redundant lines should be eliminated from a least
squares adjustment result from surveying according to the scheme previously
described for Fig. 7.5. For example, if receivers were placed on points 1, 2,
3, and 4. The vectors 1–2, 2–3, and 3–4, could have accurate fixed bias
solutions while the vector 1–4 could be a float bias solution. This difference
would result from the fact that integer biases are more easily fixed on short
lines but are difficult to fix on long lines. In this case, it would be appropriate
to eliminate vector 1–4 from the adjustment.

When a kinematic survey has been performed using the leapfrog traversing technique, the traverse or geometric figure can be adjusted in a manner similar to a static survey. The same procedure should be followed to isolate bad lines, or bad elevations, or horizontal points.

7.4.4 Datum transformations

To this point only the computation of vectors and the analysis of their accuracy have been dealt with. After having successfully isolated and eliminated bad lines, the coordinates of points in the network can be computed. The first such computation is the final minimally constrained adjustment using all good vectors. The network shown in Fig. 7.5 has three horizontal control points. One of these points would be designated as fixed and one of the vertical points would be fixed in the free adjustment. If the point HCA were held fixed, the horizontal coordinates of the other two points would be determined. These (free adjusted) values should agree with the known values within a reasonable amount. If, for example, one point agreed well with the adjusted value while the other point disagreed, one could isolate the bad control point.

In most cases, the fixed control agrees quite well and the final adjustment is then performed. The constrained (fixed) adjustment results when two or more horizontal points are held fixed (also three vertical discussed in next section). In Fig. 7.5 the horizontal coordinates of HCA, HCB, and HCC would be fixed and the GPS network would be rotated and scaled to fit these values. Rotation values of 1 or 2 arcseconds and scales of up to 10 ppm are common when the fixed control has been established by triangulation, and less than 1 arcsecond and a few ppm are expected when the fixed control is established by dual frequency GPS (e.g., supernets).

Horizontal datum. The coordinates of points in a GPS network are computed in the same manner that coordinates of triangulation or traverse points. Network coordinates are, essentially, interpolated between the fixed control points using the measured vectors to apportion the values. Geodetic coordinates are thus referenced to the values of the fixed control and are therefore on the same datum as the fixed control. The proper ellipsoidal parameters should be used when performing the least squares adjustment. For example, when North American Datum 1983 (NAD-83) coordinates are used in an adjustment, the Geodetic Reference System 1980 (GRS-80) ellipsoid should be used. When the computation of coordinates for a survey on a different datum is desired, the coordinates of the fixed control and the proper ellipsoidal values must be used for the new adjustment.

Vertical datum. The computation of elevations with GPS is complicated by the fact that GPS heights are purely geometric and referenced to a surface known as the ellipsoid while elevations measured with leveling equipment are referenced to a surface known as the geoid, cf. Sect. 10.2.3. Geodesists at various universities and national geodetic organizations have developed a mathematical model that approximates the shape of the geoid (which is quite irregular). These models are based on data from satellites, astronomic observations, and gravity measurements. These geoid models have been used to compute geoid/ellipsoid separation values (i.e., geoidal heights) on regular grids. For example, the NGS distributes a file containing these geoidal heights for every three minutes of arc in latitude and longitude for the conterminous United States. These theoretical geoidal heights can be used to provide (theoretically) more accurate adjusted elevations.

In general, the ellipsoid and geoid are separated from one another (e.g., some 30 m in U.S.) and are also tilted to one another. Over a moderate size area (10 km × 10 km) with smooth topography, one can simply assume that the geoid has a similar curvature as the ellipsoid and compute the rotation angles and translation distance between the two surfaces, cf. Sect. 10.3.3.

In practice, more than the minimum three required elevations may be available for points in a GPS network and a least squares adjustment could be performed. Referring to Fig. 7.5, the elevations of points 1, 6 and 13 have been determined and a vector has been measured (twice) to BMB, so that there are four known elevations connected by a network. The transformation of ellipsoidal heights to elevations is accomplished by holding the vertical component of the four points fixed in the adjustment. The adjustment will rotate and translate the entire network so that the elevations of all points are quite close to the true values. This technique works well where the geoid is similar in shape to the ellipsoid; however, in mountainous areas, this assumption is not valid and the difference must be accounted for.

In certain cases, many more than the minimum required elevations may be available for points in a GPS network. For example, a 100 point network covering a county sized area may have 10 elevations scattered throughout the area. Out of the 10 known elevations, there may be as many as five or six combinations of three points that satisfy the "well distributed geometrically" criterion. Therefore, to check for badly known elevations, five to six preliminary adjustments of the 100 station network would have to be performed. Another way of isolating bad elevations can be found in Collins (1989) where an example is given. This method first involves performing a minimally constrained or "free" adjustment holding one of the vertical points fixed (and at least one horizontal point). The geoidal heights are included in this free adjustment so that all points in the network now have so-called pseudoelevations. For example, in Fig. 7.5 the elevation of BMD could be fixed and the pseudoelevations of the remaining points computed by using the geoidal height differences (derived from the geoid model) and the ellipsoidal height differences (determined from the adjustment). The pseudoelevations approximate the true elevations although they contain a systematic error. In a moderate size nonmountainous area (10 km × 10 km), the pseudoelevations are normally within 10 to 20 cm of the true elevations. Many times the pseudoelevations will be within a few centimeters of the true elevations. The rotation angles can now be computed separately by a simple "regression" program. A convenient way to write such a program is to use a spread sheet program (e.g., Lotus 123). A list of each point's plane coordinates and pseudoelevation is listed in the first columns and rows of the spread sheet. The difference between the pseudoelevation and true elevation is placed in another column and zeroes are placed in the cells of this column for points not being held fixed. The normal equation matrix can now be formed automatically by ignoring data in rows where zeroes appear, using only the coordinates and height difference in rows where differences between the true and the pseudoelevations are given. Standard regression or least squares adjustment formulas are employed to compute the 3 × 3 normal equation matrix used to determine the transformation parameters. The five or six combinations of various elevations can easily be computed in a few minutes using this technique. Bad elevations can be quickly isolated in this way since the rotated elevations for the redundant points should agree within a few centimeters of the fixed elevation. After all given elevations have been checked for blunders, the final elevations of points in the network are computed. This is easily accomplished by fixing a minimum of three (geometrically diverse) elevations and performing a fixed adjustment. For example, the elevations at points 1, 6, and 18 in Fig. 7.5 would be fixed in the adjustment, and the elevation of point 13 would serve as a check. When three elevations are fixed in the adjustment, the rotation angles about the north-south and east-west axes are computed. These rotation angles serve as a further check since they should not (normally) exceed a few (e.g., 5) arcseconds.

The differences of theoretical geoidal heights are also useful in determining elevations in a small area where only one benchmark is available. The pseudoelevations for the area will be closer to the true values than if no geoid model were used and a single point were held fixed. This method should only be used when approximate elevations are desired and the technique should be used with caution since all elevations are dependent on the single fixed point.

No definite answer can be given to the question of the accuracy of GPS derived elevations since the accuracy depends upon the shape of the geoid in the particular area. Normally, in flat continental areas the geoid is relatively smooth and elevations within a 10 km × 10 km area can be determined to 3 cm or better. In mountainous areas the interpolated (holding three elevations fixed) elevations can be in error by several decimeters. In Europe where geoid models are more accurate, the errors might be one-half the errors experienced in the United States and Australia.

7.4.5   Computation of plane coordinates

GPS surveys performed for local engineering projects or for mapping normally require that plane coordinates be computed for all points. There are several conformal projections used for this purpose. The two most popular in the U.S. are the Transverse Mercator projection and the Lambert projection. In the U.S. each state has a different projection system (in some cases several) and programs to transform latitude and longitude to and from these projections can be purchased from the NGS. Also, most GPS software includes these transformation programs. In Europe, transformation programs are readily available from the various mapping and surveying organizations. Also, these transformations are included in most land surveying software packages.

It should be remembered that geodetic latitude and longitude are specifically referenced to a given datum. In the U.S., the older North American Datum 1927 (NAD-27) has a specific set of projections to use, while the NAD-83 uses a completely different set of projections. The projections between datums should never be mixed.

There exists a number of transformation programs commercially available that transform both ellipsoidal and plane coordinates between datums and projections. This software is often part of Computer Aided Design (CAD) graphics package. It should be noted that these transformation programs are not intended for geodetic use; they only approximate the coordinates of the new datum. Such programs can produce coordinates with errors of up to a meter.

Special local transformations can provide quite accurate coordinates of points that do not have published values in the new system. For example, not all geodetic points have published NAD-83 values. Software such as the NGS program LEFTI use the old and new geodetic coordinates of points surrounding the old (without new values) point to compute local transformation parameters used to compute new coordinates for the old point. This type of program works quite well and provides geodetic quality coordinates.

7.5 Survey report

A final survey report is helpful to others in analyzing the conduct of a survey. In effect, a project report should document the survey so that another competent individual could recreate all computations arriving at similar answers to those obtained. This report serves as an "audit trail" for the survey. A project report should address the following topics:

1. Location of the survey and a description of the project area. A general map showing the locality is recommended.

2. Purpose of the survey and the extent that the requirements were satisfied. Also, the specifications followed should be mentioned (e.g., FGCC GPS Specifications Order A).

3. A description of the monumentation used. It should be specifically noted if underground as well as surface marks were used. A section of the report should explain which existing monuments were searched for and which ones were found. For monuments not found, an estimate of the time spent searching for the mark should be given to aid others who wish to recover that mark. A list of all control searched for, control found, and control used is helpful in analyzing the project.

4. A description of the instrumentation used should include both the GPS equipment and conventional equipment along with serial numbers. An explanation of how the tribrachs or bipods were tested for plumb should be given. If survey towers or special range poles were used, their use should be described with an explanation of how the antenna was collimated.

5. The computation scheme for the project should be described including which version of the processing software was used and which least squares adjustment was applied. An exhibit such as Table 7.5 is particularly useful in visualizing how the survey was conducted. The satellites tracked during each session should be included in this listing or in a separate exhibit. Also, an exhibit such as Table 7.6 should be included to show the quality factors for each line. The inclusion of abstracts of each vector would also be helpful.

6. The computation of coordinates for all eccentric points should be included. In the case of horizontal points, sketches and the direct computation results should be shown. For vertical points, a copy of the level records should be included in an appendix.

7. All problems encountered should be discussed and equipment failures listed. Unusual solar activity should be mentioned as well as multipath problems and other factors affecting the survey.

8. The following lists should be included in the report:

   (a) List of loop closures.

   (b) Occupation schedule (e.g., Table 7.5).

   (c) Vector statistics (e.g., Table 7.6).

   (d) Free least squares adjustment.

   (e) Portion of fixed adjustment showing rotation angles and statistics.

   (f) List of adjusted positions and plane coordinates.

   (g) Project statistics.

   (h) Copies of "original" site occupation logs.

   (i) Equipment malfunction log.

   (j) A project sketch showing all points and control with a title box, scale, and projection tic marks.

9. Finally, a copy of the original observations or the original observations translated to the receiver independent RINEX format, cf. Sect. 9.1, should be transmitted as part of the survey. For convenience, copies of the adjustment input file and vector output files would be advantageous.

When surveys are properly performed and documented, they provide a lasting contribution to the profession. Often, data can be used in later years by others to study a particular phenomenon or the work may be included in a larger project. Proper use of measurements can only be made when the survey is thoroughly documented for posterity.

8. Mathematical models for positioning

8.1 Point positioning

8.1.1 Point positioning with code ranges

The code pseudorange at an epoch t can be modeled, cf. Eq. (6.2), by

$$R_i^j(t) = \varrho_i^j(t) + c\,\Delta\delta_i^j(t). \tag{8.1}$$

Here, $R_i^j(t)$ is the measured code pseudorange between the observing site i and the satellite j, $\varrho_i^j(t)$ is the geometric distance between the satellite and the observing point, and c is the speed of light. The last item to be explained is $\Delta\delta_i^j(t)$. This clock bias represents the combined clock offsets of the satellite and the receiver clock with respect to GPS time, cf. Eq. (6.1).

Examining Eq. (8.1), the desired point coordinates to be determined are implicit in the distance $\varrho_i^j(t)$, which can explicitly be written as

$$\varrho_i^j(t) = \sqrt{(X^j(t) - X_i)^2 + (Y^j(t) - Y_i)^2 + (Z^j(t) - Z_i)^2} \tag{8.2}$$

where $X^j(t)$, $Y^j(t)$, $Z^j(t)$ are the components of the geocentric position vector of the satellite for epoch t and X_i, Y_i, Z_i are the three unknown ECEF coordinates of the observing site. Now, the clock bias $\Delta\delta_i^j(t)$ must be investigated more detailed. For the moment consider a single epoch; a single position i is automatically implied. Each satellite contributes one unknown clock bias which can be recognized from the superscript j at the clock term. Neglecting, for the present, the site i clock bias, the pseudorange equation for the first satellite would have four unknowns. These are the three site coordinates and one clock bias of this satellite. Each additional satellite adds one equation with the same site coordinates but with a new satellite clock bias. Thus, there would always be more unknowns than measurements. Even when an additional epoch is considered, new satellite clock biases must be modeled due to clock drift. Fortunately, the information on the satellite clocks is known and transmitted via the broadcast navigation message in form of three polynomial coefficients a_0, a_1, a_2 with a reference time t_0, cf. van Dierendonck et al. (1980). Therefore, the equation

$$\delta^j(t) = a_0 + a_1(t - t_0) + a_2(t - t_0)^2 \tag{8.3}$$

enables the calculation of the satellite clock bias for epoch t. It should be noted that the polynomial (8.3) removes a great deal of the satellite clock bias but a small amount of error remains.

The combined bias term $\Delta \delta_i^j(t)$ is split into two parts by

$$\Delta \delta_i^j(t) = \delta^j(t) - \delta_i(t) \tag{8.4}$$

where the satellite related part is known by (8,3) and the receiver related term $\delta_i(t)$ remains unknown. Substituting (8.4) into (8.1) yields

$$R_i^j(t) = \varrho_i^j(t) + c\,\delta^j(t) - c\,\delta_i(t) \tag{8.5}$$

which is the equation that will be investigated with respect to the unknowns. Considering a single epoch t, there are four unknowns. These are three site coordinates and one unknown receiver clock bias $\delta_i(t)$. As mentioned above, the satellite dependent term $\delta^j(t)$ can be determined by polynomial (8.3). The four unknowns can be computed immediately if four satellites are observed simultaneously.

Equation (8.5) is now studied in a more general way. Denoting the number of satellites by n_j, and the number of epochs by n_t, then $n_j\,n_t$ observation equations are available. To obtain a unique solution, the number of unknowns must not be greater than the number of observations. Rearranging (8.5) slightly by shifting the computed satellite clock bias to the left-hand side of the equation, one gets

$$
\begin{aligned}
R_i^j(t) - c\,\delta^j(t) \;\; &= \;\; \varrho_i^j(t) \;\; - \;\; c\,\delta_i(t) \\
n_j\,n_t \qquad\; &\geq \quad 3 \quad + \quad n_t\,.
\end{aligned}
\tag{8.6}
$$

There the number of measurements and the number of unknowns are shown below each corresponding term. Limiting the relation above to one epoch (i.e., $n_t = 1$) gives $n_j \geq 4$. Note that each epoch can be considered separately and the position of site i and the clock biases for each corresponding epoch can be computed if simultaneous observations are performed. Therefore, movement of receiver i is permitted. According to the definitions in Sect. 7.1.1, this is the case in kinematic applications, where the position of a moving vehicle at any arbitrary epoch is desired. This trajectory determination or navigation is the original goal of GPS. For each epoch, the position (and velocity) of i can be determined instantaneously when at least four satellites are in view.

In static applications where site i is stationary during the observation period, the situation is slightly different. In this case, in principle, the simultaneous observation of four satellites is not necessary. This can be seen from Eq. (8.6) if two satellites (i.e., $n_j = 2$) are considered. The corresponding result $n_t \geq 3$ means that the simultaneous observation of two satellites over three epochs would theoretically suffice. In practice, however, this situation would give bad results or the computation would fail because of an ill-conditioned system of observation equations unless the epochs were widely spaced (e.g., hours). However, if observations of three epochs for two satellites are made, followed by three additional epochs (e.g., seconds apart) for two other satellites, then a solution is possible. Such an application will be rare but is imaginable under special circumstances (e.g., in urban areas).

8.1.2 Point positioning with carrier phases

Pseudoranges can also be derived from carrier phase measurements. The mathematical model for these measurements, cf. Eq. (6.9), is

$$\Phi_i^j(t) = \frac{1}{\lambda} \varrho_i^j(t) + N_i^j + f^j \, \Delta\delta_i^j(t) \, . \tag{8.7}$$

Here, $\Phi_i^j(t)$ is the measured carrier phase in cycles, λ is the wavelength, and $\varrho_i^j(t)$ is the same as for the code range model. The time independent phase ambiguity N_i^j is an integer number and, therefore, often called the integer ambiguity or integer unknown. The term f^j denotes the frequency of the satellite signal in cycles per second.

If Eq. (8.4) is substituted into Eq. (8.7), the phase model becomes

$$\Phi_i^j(t) = \frac{1}{\lambda} \varrho_i^j(t) + N_i^j + f^j \, \delta^j(t) - f^j \, \delta_i(t) \, . \tag{8.8}$$

The satellite clock bias $\delta^j(t)$ is again assumed to be known, cf. Eq. (8.3). For n_j satellites, n_t epochs, and the single site i, there are $n_j n_t$ possible measurements. The number of unknowns are listed below the appropriate terms on the right-hand side of the following equation:

$$\Phi_i^j(t) - f^j \, \delta^j(t) \quad = \quad \frac{1}{\lambda} \varrho_i^j(t) \quad + \quad N_i^j \quad - \quad f^j \, \delta_i(t)$$
$$n_j \, n_t \qquad\qquad \geq \qquad 3 \qquad + \quad n_j \quad + \qquad n_t \, . \tag{8.9}$$

A solution for a single epoch (i.e., $n_t = 1$) is only possible when the n_j integer ambiguities are disregarded. In this case, the phase range model is equivalent to the code range model and it follows $n_j \geq 4$. This essentially means that the phase model may also be used for kinematic applications provided that the ambiguities have been resolved by initial (e.g., static) observations. When the ambiguities are to be determined, theoretically the minimum number of satellites is $n_j = 2$. This requires a minimum of $n_t = 5$ observation epochs. Practically, $n_j = 4$ satellites would give reasonable results. In this case, the required number of epochs is $n_t \geq 3$.

8.1.3 Point positioning with Doppler data

The mathematical model for Doppler data, cf. Eq. (6.10), is

$$D_i^j(t) = \dot{\varrho}_i^j(t) + c\,\Delta\dot{\delta}_i^j(t) \tag{8.10}$$

and can be considered as time derivative of a code or phase pseudorange. In this equation, $D_i^j(t)$ denotes the observed Doppler shift scaled to range rate, $\dot{\varrho}_i^j(t)$ is the instantaneous radial velocity between the satellite and the receiver, and $\Delta\dot{\delta}_i^j(t)$ is the time derivative of the combined clock bias term. The radial velocity for a stationary receiver, cf. Eq. (4.48),

$$\dot{\varrho}_i^j(t) = \frac{\underline{\varrho}^j(t) - \underline{\varrho}_i}{\|\underline{\varrho}^j(t) - \underline{\varrho}_i\|} \cdot \underline{\dot{\varrho}}^j(t) \tag{8.11}$$

relates the unknown position vector $\underline{\varrho}_i$ of the receiver with the instantaneous position vector $\underline{\varrho}^j(t)$ and velocity vector $\underline{\dot{\varrho}}^j(t)$ of the satellite. These vectors can be calculated from the satellite's ephemerides. The contribution of the satellite clock to $\Delta\dot{\delta}_i^j(t)$ is given by, cf. Eq. (8.3),

$$\dot{\delta}^j(t) = a_1 + 2a_2(t - t_0) \tag{8.12}$$

and is known. Summarizing, one can state that the observation equation (8.10) contains four unknowns. These unknowns are the three receiver coordinates $\underline{\varrho}_i$ and the receiver clock drift $\dot{\delta}_i(t)$. Hence, compared to the code range model, the Doppler equations contain only the receiver clock drift instead of the receiver clock offset. Static point positioning is thus possible.

The concept of a combined code pseudorange and Doppler data processing for static positioning is stressed by Ashjaee et al. (1989). In this case there is a total of five unknowns. These unknowns are three point coordinates, the receiver clock offset, and the receiver clock drift. Each satellite contributes two equations, one code pseudorange and one Doppler equation. Therefore, three satellites are sufficient to solve for the five unknowns.

The similarity of the pseudorange and the Doppler equation gives rise to the question of a linear dependence of the equations. However, it can be shown that the lines of constant pseudoranges and the lines of constant Doppler are orthogonal, cf. Jorgensen (1980), Hatch (1982), Ashjaee et al. (1989). The pseudoranges are measured along the line of sight between receiver and satellite, and the Doppler, according to Jorgensen (1980), "provides information along the direction of the component of relative velocity that is perpendicular to the line of sight". Thus, pseudorange and Doppler equations are independent.

8.2 Relative positioning

The objective of relative positioning is to determine the coordinates of an unknown point with respect to a known point which for most applications is stationary. In other words, relative positioning aims at the determination of the vector between the two points which is often called a "baseline". In Fig. 8.1 A denotes the (known) reference point, B the unknown point, and $\underline{b}_{AB}$ the baseline vector. Introducing the corresponding position vectors $\underline{X}_A$, $\underline{X}_B$, the relation

$$\underline{X}_B = \underline{X}_A + \underline{b}_{AB} \tag{8.13}$$

may be formulated and the components of the baseline vector $\underline{b}_{AB}$ are

$$\underline{b}_{AB} = \begin{bmatrix} X_B - X_A \\ Y_B - Y_A \\ Z_B - Z_A \end{bmatrix} = \begin{bmatrix} \Delta X_{AB} \\ \Delta Y_{AB} \\ \Delta Z_{AB} \end{bmatrix} . \tag{8.14}$$

The mathematical models for the code range and phase range, cf. Eqs. (8.6) and (8.9), respectively, may be applied analogously where the only difference is the inclusion of the known coordinates of the reference point. These coordinates must be given in the WGS-84 system and are usually approximated by a code range solution. As a rule of thumb, an accuracy of about 50 m is sufficient for all practical purposes.

Relative positioning is most effective if simultaneous observations are made at both the reference and the unknown points. Simultaneity means that the observation time tags for the two points are the same. Assuming such simultaneous observations at the two points A and B to satellites j and k, linear combinations can be formed leading to single-differences, double-differences, and triple-differences. Most postprocessing software uses these three difference techniques, so their basic mathematical modeling is shown in the following sections.

8.2.1 Phase differences

Single-differences. Two points and one satellite are involved. Denoting the

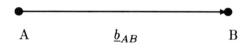

Fig. 8.1. Relative positioning

points by A and B and the satellite by j and using Eq. (8.9), the phase equations for the two points are

$$\Phi_A^j(t) - f^j \, \delta^j(t) = \frac{1}{\lambda} \varrho_A^j(t) + N_A^j - f^j \, \delta_A(t)$$

$$\Phi_B^j(t) - f^j \, \delta^j(t) = \frac{1}{\lambda} \varrho_B^j(t) + N_B^j - f^j \, \delta_B(t)$$

(8.15)

and the difference of the two equations is

$$\Phi_B^j(t) - \Phi_A^j(t) = \frac{1}{\lambda} [\varrho_B^j(t) - \varrho_A^j(t)] + N_B^j - N_A^j$$
$$- f^j [\delta_B(t) - \delta_A(t)] .$$

(8.16)

Equation (8.16) is referred to as the single-difference equation. This equation stresses one aspect of the solution for the unknowns on the right-hand side. A system of such equations would lead to a rank deficiency even in the case of an arbitrarily large redundancy. This can be seen from the coefficients of the ambiguities and of the receiver clock biases. In both cases, the absolute values of the coefficients for the two points are the same. This means that the adjustment design matrix has linearly dependent columns and a rank deficiency exists. Therefore, the relative quantities

$$N_{AB}^j = N_B^j - N_A^j$$

$$\delta_{AB}(t) = \delta_B(t) - \delta_A(t)$$

(8.17)

are introduced. Using additionally the shorthand notations

$$\Phi_{AB}^j(t) = \Phi_B^j(t) - \Phi_A^j(t)$$

$$\varrho_{AB}^j(t) = \varrho_B^j(t) - \varrho_A^j(t) ,$$

(8.18)

and substituting (8.17) and (8.18) into (8.16),

$$\Phi_{AB}^j(t) = \frac{1}{\lambda} \varrho_{AB}^j(t) + N_{AB}^j - f^j \, \delta_{AB}(t)$$

(8.19)

is the final form of the single-difference equation. Compared to the phase equation (8.9), the satellite clock bias has canceled.

Double-differences. Assuming the two points A, B, and the two satellites j, k to be involved, two single-differences according to Eq. (8.19) may be formed:

$$\Phi_{AB}^j(t) = \frac{1}{\lambda} \varrho_{AB}^j(t) + N_{AB}^j - f^j \, \delta_{AB}(t)$$

$$\Phi_{AB}^k(t) = \frac{1}{\lambda} \varrho_{AB}^k(t) + N_{AB}^k - f^k \, \delta_{AB}(t) .$$

(8.20)

To get a double-difference, these single-differences are subtracted. Assuming equal frequencies $f^j = f^k$, the result is

$$\Phi^k_{AB}(t) - \Phi^j_{AB}(t) = \frac{1}{\lambda}\left[\varrho^k_{AB}(t) - \varrho^j_{AB}(t)\right] + N^k_{AB} - N^j_{AB}. \quad (8.21)$$

Using in analogous way to (8.18) shorthand notations for the satellites j and k, the final form of the double-difference equation is

$$\Phi^{jk}_{AB}(t) = \frac{1}{\lambda}\varrho^{jk}_{AB}(t) + N^{jk}_{AB}. \quad (8.22)$$

The canceling effect of the receiver clock biases is the reason why double-differences are used. This cancellation resulted from the assumptions of simultaneous observations and equal frequencies of the satellite signals.

Symbolically, the convention

$$*^{jk}_{AB} = *^k_{AB} - *^j_{AB} \quad (8.23)$$

has been introduced where the asterisk may be replaced by Φ, ϱ, or N. Note that because of (8.17) and (8.18) these terms comprising two subscripts and two superscripts are actually composed of four terms. The symbolic notation

$$*^{jk}_{AB} = *^k_B - *^j_B - *^k_A + *^j_A \quad (8.24)$$

characterizes in detail the terms in the double-difference equation:

$$\begin{aligned}
\Phi^{jk}_{AB}(t) &= \Phi^k_B(t) - \Phi^j_B(t) - \Phi^k_A(t) + \Phi^j_A(t) \\
\varrho^{jk}_{AB}(t) &= \varrho^k_B(t) - \varrho^j_B(t) - \varrho^k_A(t) + \varrho^j_A(t) \\
N^{jk}_{AB} &= N^k_B \quad - N^j_B \quad - N^k_A \quad + N^j_A.
\end{aligned} \quad (8.25)$$

Triple-differences. So far only one epoch t has been considered. To eliminate the time independent ambiguities, Remondi (1984) has suggested to difference double-differences between two epochs. Denoting the two epochs by t_1 and t_2, then

$$\begin{aligned}
\Phi^{jk}_{AB}(t_1) &= \frac{1}{\lambda}\,\varrho^{jk}_{AB}(t_1) + N^{jk}_{AB} \\
\Phi^{jk}_{AB}(t_2) &= \frac{1}{\lambda}\,\varrho^{jk}_{AB}(t_2) + N^{jk}_{AB}
\end{aligned} \quad (8.26)$$

are the two double-differences and

$$\Phi^{jk}_{AB}(t_2) - \Phi^{jk}_{AB}(t_1) = \frac{1}{\lambda}\left[\varrho^{jk}_{AB}(t_2) - \varrho^{jk}_{AB}(t_1)\right] \quad (8.27)$$

is the triple-difference which may be written in the more simplified form

$$\Phi_{AB}^{jk}(t_{12}) = \frac{1}{\lambda}\, \varrho_{AB}^{jk}(t_{12}) \tag{8.28}$$

if the symbolic formula

$$*(t_{12}) = *(t_2) - *(t_1) \tag{8.29}$$

is applied to the terms Φ and ϱ. It should be noted that both $\Phi_{AB}^{jk}(t_{12})$ and $\varrho_{AB}^{jk}(t_{12})$ are actually composed of eight terms each. Resubstituting (8.27) and either (8.24) or (8.25), yields

$$\begin{aligned}\Phi_{AB}^{jk}(t_{12}) = \quad &\Phi_B^k(t_2) - \Phi_B^j(t_2) - \Phi_A^k(t_2) + \Phi_A^j(t_2) \\ &- \Phi_B^k(t_1) + \Phi_B^j(t_1) + \Phi_A^k(t_1) - \Phi_A^j(t_1)\end{aligned} \tag{8.30}$$

and

$$\begin{aligned}\varrho_{AB}^{jk}(t_{12}) = \quad &\varrho_B^k(t_2) - \varrho_B^j(t_2) - \varrho_A^k(t_2) + \varrho_A^j(t_2) \\ &- \varrho_B^k(t_1) + \varrho_B^j(t_1) + \varrho_A^k(t_1) - \varrho_A^j(t_1)\,.\end{aligned} \tag{8.31}$$

There are two main advantages of the triple-differences which are related. These are the canceling effect for the ambiguities and thus the immunity of the triple-differences to changes in the ambiguities. Such changes are called cycle slips and are treated in more detail in Sect. 9.1.2.

8.2.2   Correlations of the phase combinations

In general, there are two groups of correlations, (1) the physical, and (2) the mathematical correlations. The phases from one satellite received at two points, for example $\Phi_A^j(t)$ and $\Phi_B^j(t)$, are physically correlated since they refer to the same satellite. Usually, the physical correlation is not taken into account. Therefore, main interest is directed to the mathematical correlations introduced by differencing.

The assumption may be made that the phase errors show a random behaviour resulting in a normal distribution with expectation value zero and variance σ^2. Measured (or raw) phases are, therefore, linearly independent or uncorrelated. Introducing a vector $\underline{\Phi}$ containing the phases, then

$$\text{cov}(\underline{\Phi}) = \sigma^2\, \underline{I} \tag{8.32}$$

is the covariance matrix for the phases where $\underline{I}$ is the unit matrix.

Single-differences. Considering the two points A, B and the satellite j at epoch t gives

$$\Phi_{AB}^j(t) = \Phi_B^j(t) - \Phi_A^j(t) \tag{8.33}$$

as the corresponding single-difference. Forming a second single-difference for the same two points but with another satellite k for the same epoch, yields

$$\Phi^k_{AB}(t) = \Phi^k_B(t) - \Phi^k_A(t). \tag{8.34}$$

The two single-differences may be computed from the matrix-vector relation

$$\underline{SD} = \underline{C}\,\underline{\Phi} \tag{8.35}$$

where

$$\underline{SD} = \left[\begin{array}{c} \Phi^j_{AB}(t) \\ \Phi^k_{AB}(t) \end{array}\right]$$

$$\underline{C} = \left[\begin{array}{cccc} -1 & 1 & 0 & 0 \\ 0 & 0 & -1 & 1 \end{array}\right] \qquad \underline{\Phi} = \left[\begin{array}{c} \Phi^j_A(t) \\ \Phi^j_B(t) \\ \Phi^k_A(t) \\ \Phi^k_B(t) \end{array}\right]. \tag{8.36}$$

The covariance law applied to Eq. (8.35) yields

$$\mathrm{cov}(\underline{SD}) = \underline{C}\,\mathrm{cov}(\underline{\Phi})\,\underline{C}^T \tag{8.37}$$

and, by substituting Eq. (8.32),

$$\mathrm{cov}(\underline{SD}) = \underline{C}\,\sigma^2\,\underline{I}\,\underline{C}^T = \sigma^2\,\underline{C}\,\underline{C}^T \tag{8.38}$$

is obtained for the covariance of the single-differences. Taking $\underline{C}$ from (8.36), the matrix product

$$\underline{C}\,\underline{C}^T = 2\left[\begin{array}{cc} 1 & 0 \\ 0 & 1 \end{array}\right] = 2\,\underline{I} \tag{8.39}$$

may be substituted into (8.38) leading to

$$\mathrm{cov}(\underline{SD}) = 2\sigma^2\,\underline{I}. \tag{8.40}$$

This shows that single-differences are uncorrelated. Note that the dimension of the unit matrix in (8.40) corresponds to the number of single-differences at epoch t whereas the factor 2 does not depend on the number of single-differences. Considering more than one epoch, the covariance matrix is again a unit matrix with the dimension equivalent to the total number of single-differences.

Double-differences. Now, three satellites j, k, ℓ with j as reference satellite are considered. For the two points A, B, and epoch t the double-differences

$$\Phi^{jk}_{AB}(t) = \Phi^k_{AB}(t) - \Phi^j_{AB}(t)$$

$$\Phi^{j\ell}_{AB}(t) = \Phi^\ell_{AB}(t) - \Phi^j_{AB}(t) \tag{8.41}$$

can be derived from the single-differences. These two equations can be written in the matrix-vector form

$$\underline{DD} = \underline{C}\,\underline{SD} \tag{8.42}$$

where

$$\underline{DD} = \begin{bmatrix} \Phi^{jk}_{AB}(t) \\ \Phi^{j\ell}_{AB}(t) \end{bmatrix}$$

$$\underline{C} = \begin{bmatrix} -1 & 1 & 0 \\ -1 & 0 & 1 \end{bmatrix} \qquad \underline{SD} = \begin{bmatrix} \Phi^{j}_{AB}(t) \\ \Phi^{k}_{AB}(t) \\ \Phi^{\ell}_{AB}(t) \end{bmatrix} \tag{8.43}$$

have been introduced. The covariance matrix for the double-differences is given by

$$\mathrm{cov}(\underline{DD}) = \underline{C}\,\mathrm{cov}(\underline{SD})\,\underline{C}^T \tag{8.44}$$

and substituting (8.40) leads to

$$\mathrm{cov}(\underline{DD}) = 2\sigma^2\,\underline{C}\,\underline{C}^T \tag{8.45}$$

or, explicitly, using $\underline{C}$ from (8.43)

$$\mathrm{cov}(\underline{DD}) = 2\sigma^2 \begin{bmatrix} 2 & 1 \\ 1 & 2 \end{bmatrix}. \tag{8.46}$$

This shows that the double-differences are correlated. The weight or correlation matrix $\underline{P}(t)$ is obtained from the inverse of the covariance matrix

$$\underline{P}(t) = [\mathrm{cov}(\underline{DD})]^{-1} = \frac{1}{2\sigma^2}\frac{1}{3} \begin{bmatrix} 2 & -1 \\ -1 & 2 \end{bmatrix} \tag{8.47}$$

where two double-differences at one epoch were used. Generally, with n_{DD} being the number of double-differences at epoch t, the correlation matrix is given by

$$\underline{P}(t) = \frac{1}{2\sigma^2}\frac{1}{n_{DD}+1} \begin{bmatrix} n_{DD} & -1 & -1 & \cdots \\ -1 & n_{DD} & -1 & \cdots \\ -1 & & & \\ \vdots & \cdots & & n_{DD} \end{bmatrix} \tag{8.48}$$

where the dimension of the matrix is $n_{DD} \times n_{DD}$. For a better illustration, assume four double-differences. In this case the correlation matrix is the 4×4 matrix

$$\underline{P}(t) = \frac{1}{2\sigma^2} \frac{1}{5} \begin{bmatrix} 4 & -1 & -1 & -1 \\ -1 & 4 & -1 & -1 \\ -1 & -1 & 4 & -1 \\ -1 & -1 & -1 & 4 \end{bmatrix}. \tag{8.49}$$

So far only one epoch has been considered. For epochs $t_1, t_2, t_3, \ldots$ the correlation matrix becomes a block-diagonal matrix

$$\underline{P}(t) = \begin{bmatrix} \underline{P}(t_1) & & & \\ & \underline{P}(t_2) & & \\ & & \underline{P}(t_3) & \\ & & & \ddots \end{bmatrix} \tag{8.50}$$

where each "element" of the matrix is itself a matrix. The matrices $\underline{P}(t_1)$, $\underline{P}(t_2)$, $\underline{P}(t_3), \ldots$ do not necessarily have to be of the same dimension because there may be different numbers of double-differences at different epochs.

Triple-differences. The triple-difference equations are slightly more complicated because several different cases must be considered. The covariance of a single triple-difference is computed by applying the covariance propagation law to the relation, cf. Eqs. (8.30) and (8.33),

$$\Phi_{AB}^{jk}(t_{12}) = \Phi_{AB}^{k}(t_2) - \Phi_{AB}^{j}(t_2) - \Phi_{AB}^{k}(t_1) + \Phi_{AB}^{j}(t_1). \tag{8.51}$$

Now, two triple-differences with the same epochs and sharing one satellite are considered. The first triple-difference using the satellites j, k is given by Eq. (8.51). The second triple-difference corresponds to the satellites j, ℓ:

$$\Phi_{AB}^{jk}(t_{12}) = \Phi_{AB}^{k}(t_2) - \Phi_{AB}^{j}(t_2) - \Phi_{AB}^{k}(t_1) + \Phi_{AB}^{j}(t_1)$$

$$\Phi_{AB}^{j\ell}(t_{12}) = \Phi_{AB}^{\ell}(t_2) - \Phi_{AB}^{j}(t_2) - \Phi_{AB}^{\ell}(t_1) + \Phi_{AB}^{j}(t_1). \tag{8.52}$$

By introducing

$$\underline{TD} = \begin{bmatrix} \Phi_{AB}^{jk}(t_{12}) \\ \Phi_{AB}^{j\ell}(t_{12}) \end{bmatrix}$$

$$\underline{C} = \begin{bmatrix} 1 & -1 & 0 & -1 & 1 & 0 \\ 1 & 0 & -1 & -1 & 0 & 1 \end{bmatrix} \quad \underline{SD} = \begin{bmatrix} \Phi_{AB}^{j}(t_1) \\ \Phi_{AB}^{k}(t_1) \\ \Phi_{AB}^{\ell}(t_1) \\ \Phi_{AB}^{j}(t_2) \\ \Phi_{AB}^{k}(t_2) \\ \Phi_{AB}^{\ell}(t_2) \end{bmatrix} \tag{8.53}$$

the vector-matrix relation

$$TD = C\ SD \tag{8.54}$$

can be formed and the covariance for the triple-difference follows from

$$\text{cov}(\underline{TD}) = \underline{C}\,\text{cov}(\underline{SD})\,\underline{C}^T \tag{8.55}$$

or, by substituting (8.40),

$$\text{cov}(\underline{TD}) = 2\sigma^2\,\underline{C}\,\underline{C}^T \tag{8.56}$$

is obtained which, using (8.53), yields

$$\text{cov}(\underline{TD}) = 2\sigma^2 \begin{bmatrix} 4 & 2 \\ 2 & 4 \end{bmatrix} \tag{8.57}$$

for the two triple-differences (8.52). The tedious derivation may be abbreviated by setting up the following symbolic table

Epoch	t_1			t_2		
SD for Sat	j	k	ℓ	j	k	ℓ
$TD^{jk}(t_{12})$	1	-1	0	-1	1	0
$TD^{j\ell}(t_{12})$	1	0	-1	-1	0	1

$$\tag{8.58}$$

where the point names A, B have been omitted. It can be seen that the triple-difference $TD^{jk}(t_{12})$ for example is composed of the two single-differences (with the signs according to the table) for the satellites j and k at epoch t_1 and of the two single-differences for the same satellites but epoch t_2. Accordingly, the same is true for the other triple-difference $TD^{j\ell}(t_{12})$. Thus, the coefficients of (8.58) are the same as those of matrix $\underline{C}$ in Eq. (8.53). Finally, the product $\underline{C}\,\underline{C}^T$, appearing in Eq. (8.56), is also aided by referring to the table above. All combinations of inner products of the two rows (one row represents one triple-difference) must be taken. The inner product (row 1 · row 1) yields the first-row, first-column element of $\underline{C}\,\underline{C}^T$, the inner product (row 1 · row 2) yields the first-row, second-column element of $\underline{C}\,\underline{C}^T$, etc. Based on the general formula (8.51) and the table (8.58), arbitrary cases may be derived easily and systematically. The subsequent diagram shows the second group of triple-difference correlations if adjacent epochs t_1, t_2, t_3

are taken. Two cases are considered:

Epoch	t_1			t_2			t_3			$\underline{C}\,\underline{C}^T$	
SD for Sat	j	k	ℓ	j	k	ℓ	j	k	ℓ		
$TD^{jk}(t_{12})$	1	−1	0	−1	1	0	0	0	0	4	−2
$TD^{jk}(t_{23})$	0	0	0	1	−1	0	−1	1	0	−2	4
$TD^{jk}(t_{12})$	1	−1	0	−1	1	0	0	0	0	4	−1
$TD^{j\ell}(t_{23})$	0	0	0	1	0	−1	−1	0	1	−1	4

$$(8.59)$$

It can be seen from Table (8.59) that an exchange of the satellites for one triple-difference causes a change of the sign in the off-diagonal elements of the matrix $\underline{C}\,\underline{C}^T$. Therefore, the correlation of $TD^{kj}(t_{12})$ and $TD^{j\ell}(t_{23})$ produces a $+1$ as off-diagonal element. Based on a table such as (8.59), each case may be handled with ease. According to Remondi (1984), pp. 142–143, computer program adaptions require only a few simple rules.

8.2.3 Static relative positioning

In a static survey of a single baseline vector between points A and B, the two receivers must stay stationary during the entire observation session. In the following, the single-, double-, and triple-differencing are investigated with respect to the number of observation equations and unknowns. It is assumed that the two sites A and B are able to observe the same satellites at the same epochs. The practical problem of satellite blockage is not considered here. The number of epochs is again denoted by n_t, and n_j denotes the number of satellites.

The undifferenced phase as shown in Eq. (8.9) (where the satellite clock is assumed to be known) is not included here, because there would be no connection (no common unknown) between point A and point B. The two data sets could be solved separately, which would be equivalent to point positioning.

A single-difference may be expressed for each epoch and for each satellite. The number of measurements is, therefore, $n_t\,n_j$. The number of unknowns are written below the corresponding term of the single-difference equation, cf. Eq. (8.19):

$$\Phi^j_{AB}(t) = \frac{1}{\lambda}\,\varrho^j_{AB}(t) + N^j_{AB} \quad - \quad f^j\,\delta_{AB}(t)$$

$$(8.60)$$

$$n_j\,n_t \;\geq\; 3 \;+\; n_j \;+\; n_t\,.$$

The equation/unknown relationship may be rewritten as

$$n_t \geq \frac{n_j + 3}{n_j - 1} \tag{8.61}$$

to solve for the number of epochs. What are the minimum requirements theoretically? One satellite does not provide a solution because the denominator of (8.61) becomes zero. With two satellites $n_t \geq 5$ results, and for the normal case of four satellites $n_t \geq \frac{7}{3}$ or, consequently, $n_t \geq 3$ is obtained.

For double-differences the relationship of measurements and unknowns is achieved using the same logic. Note that for one double-difference two satellites are necessary. For n_j satellites therefore $n_j - 1$ double-differences are obtained at each epoch, so that the total number of double-differences is $(n_j - 1)\, n_t$. The number of unknowns is found by

$$\Phi_{AB}^{jk}(t) \quad = \frac{1}{\lambda}\, \varrho_{AB}^{jk}(t) + \quad N_{AB}^{jk}$$

$$(n_j - 1)\, n_t \geq \quad 3 \quad + (n_j - 1) \tag{8.62}$$

or

$$n_t \geq \frac{n_j + 2}{n_j - 1}. \tag{8.63}$$

Hence, the minimum number of satellites is two yielding $n_t = 4$. In the case of four satellites, a minimum of two epochs is required. To avoid linearly dependent equations when forming double-differences, a reference satellite is used, against which the measurements of the other satellites are differenced. For example, take the case where measurements are made to the satellites $6, 9, 11, 12$, and 6 is used as reference satellite. Then, at each epoch the following double-differences can be formed: (9-6), (11-6), and (12-6). Other double-differences are linear combinations and thus linearly dependent. For instance, the double-difference (11-9) can be formed by subtracting (11-6) and (9-6).

The triple-differencing mathematical model includes only the three unknown point coordinates. For a single triple-difference two epochs are necessary. Consequently, in the case of n_t epochs, $n_t - 1$ linearly independent epoch combinations are possible. Thus,

$$\Phi_{AB}^{jk}(t_{12}) \quad = \frac{1}{\lambda}\, \varrho_{AB}^{jk}(t_{12})$$

$$(n_j - 1)(n_t - 1) \geq \quad 3 \tag{8.64}$$

are the resulting equations. The equation/unknown relationship may be written as

$$n_t \geq \frac{n_j + 2}{n_j - 1} \tag{8.65}$$

which yields $n_t \geq 4$ epochs when the minimum number of satellites $n_j = 2$ is introduced. For $n_j = 4$ satellites $n_t \geq 2$ epochs are required.

This completes the discussion on static relative positioning. As shown, each of the mathematical models: single-difference, double-difference, triple-difference, may be used. The relationships between equations and unknowns will be referred to again in the discussion of the kinematic case.

8.2.4 Kinematic relative positioning

In kinematic relative positioning, the receiver on the known point A of the baseline vector remains fixed. The second receiver moves, and its position is to be determined for arbitrary epochs. The models for single-, double-, and triple-difference implicitly contain the motion in the geometric distance. Considering point B and satellite j, the geometric distance in the static case is given by, cf. (8.2),

$$\varrho_B^j(t) = \sqrt{(X^j(t) - X_B)^2 + (Y^j(t) - Y_B)^2 + (Z^j(t) - Z_B)^2} \quad (8.66)$$

and in the kinematic case by

$$\varrho_B^j(t) = \frac{}{\sqrt{(X^j(t) - X_B(t))^2 + (Y^j(t) - Y_B(t))^2 + (Z^j(t) - Z_B(t))^2}} \quad (8.67)$$

where the time dependence for point B appears. In this mathematical model, three coordinates are unknown at each epoch. Thus, the total number of unknown site coordinates is $3\,n_t$ for n_t epochs. The equation/unknown relationships for the kinematic case follow from static single-, double-, and triple-difference models, cf. Eqs. (8.60), (8.62), (8.64):

$$
\begin{array}{lll}
\text{Single-difference:} & n_j\,n_t & \geq 3\,n_t + n_j + n_t \\[4pt]
\text{Double-difference:} & (n_j - 1)\,n_t & \geq 3\,n_t + (n_j - 1) \qquad (8.68) \\[4pt]
\text{Triple-difference:} & (n_j - 1)(n_t - 1) \geq 3\,n_t\,. &
\end{array}
$$

The continuous motion of the roving receiver restricts the available data for the determination of its position to one epoch. But none of the above models provides a useful solution for $n_t = 1$. What is done to modify these models is reduce the number of unknowns by eliminating the ambiguity unknowns. Omitting these in the single- and double-difference case leads to the modified observation requirements

$$
\begin{array}{ll}
\text{Single-difference:} & n_j \geq 4 \\[4pt]
\text{Double-difference:} & n_j \geq 4
\end{array}
\qquad (8.69)
$$

for $n_t = 1$. Triple-differences could be used if the coordinates of the roving receiver were known at the reference epoch. In this case, the relationship would be $(n_j - 1)(n_t - 1) \geq 3(n_t - 1)$ which leads to

$$\text{Triple-difference:} \quad n_j \geq 4 \ . \tag{8.70}$$

Hence, all of the models end up again with the fundamental requirement of four simultaneously observable satellites.

Omitting the ambiguities for single- and double-difference means that they must be known. The corresponding equations are simply obtained by rewriting (8.60) and (8.62) with the ambiguities shifted to the left-hand side. The single-differences become

$$\Phi_{AB}^j(t) - N_{AB}^j = \frac{1}{\lambda}\, \varrho_{AB}^j(t) - f^j\, \delta_{AB}(t) \tag{8.71}$$

and the double-differences

$$\Phi_{AB}^{jk} - N_{AB}^{jk} = \frac{1}{\lambda}\, \varrho_{AB}^{jk}(t) \tag{8.72}$$

where the unknowns now show up only on the right-hand sides. For the triple-difference, reducing the number of unknown points by one presupposes one known position.

Thus, all of the equations can be solved if one position of the moving receiver is known. Preferably (but not necessarily), this will be the starting point of the moving receiver. The baseline related to this starting point is also denoted as the starting vector. With a known starting vector the ambiguities are determined and are known for all subsequent positions of the roving receiver as long as no loss of signal lock occurs and a minimum of four satellites is in view.

Static initialization. Three methods are available for the determination of the starting vector. In the first method, the moving receiver is initially placed at a known point, creating a known starting vector. The ambiguities can then be calculated from the double-difference model (8.62) as real values and are then fixed to integers. A second method is to perform a static determination of the starting vector. The third initialization technique is the antenna swap method according to B. Remondi. The antenna swap is performed as follows: (1) denote the reference mark as A and the starting position of the moving receiver as B, (2) a few measurements are taken in this configuration, and (3) with continuous tracking, the receiver at A is moved to B while the receiver at B is moved to A, (4) where again a few measurements are taken. This is sufficient to precisely determine the starting vector in a very short time (e.g., 30 seconds). Detailed explanations and formulas are given in Remondi (1986) and Hofmann-Wellenhof and Remondi (1988).

Kinematic initialization. Special applications require kinematic GPS without static initialization since the moving object whose position is to be calculated is in a permanent motion (e.g., an airplane while flying). Translated to model equations this means that the most challenging case is the determination of the ambiguities on-the-fly (OTF). The solution requires an instantaneous ambiguity resolution or an instantaneous positioning (i.e., for a single epoch). This strategy sounds very simple and it actually is. The main problem is to find the position as fast and as accurately as possible. This is achieved by starting with approximations for the position and improving them by least squares adjustments or search techniques as described in Sect. 9.1.3.

For more insight, one of the many approaches is presented here. This method was proposed by Remondi (1991a) and is an easily understood and fairly efficient method.

The double-difference

$$\Phi_{AB}^{jk}(t) = \frac{1}{\lambda} \varrho_{AB}^{jk}(t) + N_{AB}^{jk} \tag{8.73}$$

for one receiver at A and the second receiver at B for the two satellites j and k at epoch t is defined in (8.62). Since now kinematic GPS is considered, this equation is slightly reformulated to account for the motion of one of the receivers. Therefore, the receivers are numbered 1 and 2 and those terms which are purely receiver dependent get the same indices. Thus,

$$\Phi_{12}^{jk}(t) = \frac{1}{\lambda} \varrho_{AB}^{jk}(t) + N_{12}^{jk} \tag{8.74}$$

shows that the measured phases and the ambiguities are receiver dependent whereas the distances depend on the location of the site A and B. Receiver 1 is now assumed to be stationary at site A while receiver 2 moves and is assumed at site C at an epoch t_i. The double-difference equation

$$\Phi_{12}^{jk}(t_i) = \frac{1}{\lambda} \varrho_{AC}^{jk}(t_i) + N_{12}^{jk} \tag{8.75}$$

reflects this situation. In an undisturbed environment, the ambiguities are time independent and thus N_{12}^{jk} do not change during the observation. Unlike the static initialization with a known starting vector, here the position of the reference site B is unknown. Therefore, an approximate value for B must be computed and improved by a search technique. For the approximation, Remondi (1991a) proposes a carrier-smoothed code range solution.

According to Eq. (6.25), the code range at an initial epoch t can be calculated from the code range at epoch t_i by subtracting the carrier phase
difference measured between these epochs. Thus,

$$R_B^j(t) = R_B^j(t_i) - \lambda\left[\Phi_B^j(t_i) - \Phi_B^j(t)\right] + \varepsilon \qquad (8.76)$$

where a noise term ε has been added. This equation can be interpreted as a
mapping of epoch t_i to epoch t and allows an averaging (i.e., smoothing) of
the initial code range. By means of such smoothed code ranges to at least
$j = 4$ satellites the position of B is calculated as described in Sect. 8.1.1.
Afterwards a search technique supplemented by Doppler observables is initialized to improve the position of B so that the ambiguities in (8.74) can
be fixed to integer values. The integer ambiguities are then substituted into
(8.75) and enable the calculation of site C at an epoch t_i.

8.2.5 Mixed-mode relative positioning

The combination of static and kinematic leads to a noncontinuous motion.
The moving receiver stops at the points being surveyed. Hence, there are
generally measurement data of more than one epoch per point. The trajectory between the stop points is of no interest to land surveyors and the
corresponding data may be discarded if the remaining data are problem free.
Nevertheless, the tracking while in motion must not be interrupted; otherwise a new initialization (i.e., determination of a starting vector) must be
performed.

Pseudokinematic relative positioning deserves special treatment. The
notation is somehow misleading because the method is more static than
kinematic. However, the name has been chosen by its inventor and will be
used here. This method should better be called intermittent static or broken
static, cf. Remondi (1990b). The pseudokinematic technique requires that
each point to be surveyed is reoccupied. There is no requirement to maintain
lock between the two occupations. This means that, in principle, the receiver
can be turned off while moving, cf. Ashtech (1991).

The pseudokinematic observation data are collected and processed in
the same way as static observations. The data collected while the receiver
was moving are ignored. Since each site is reoccupied, the pseudokinematic method can be identified with static surveying with large data gaps,
cf. Kleusberg (1990a). The mathematical model for e.g. double-differences
corresponds to Eq. (8.62) where generally two sets of phase ambiguities
must be resolved since the point is occupied at different times. Following a proposal by Remondi (1990b), processing of the data could start with
a triple-difference solution for the few minutes of data collected during the

two occupations of a site. Based on this solution, the connection between the two ambiguity sets is computed. This technique will only work in cases where the triple-difference solution is sufficiently good; Remondi (1990b) claims roughly 30 cm. After the successful ambiguity connection the normal double-difference solutions are performed.

The time span between the two occupations is an important factor affecting accuracy. The solution depends on the change in the satellite-receiver geometry. Willis and Boucher (1990) investigate the accuracy improvements by an increasing time span between the two occupations. As a rule of thumb, the minimum time span should be one hour.

9. Data processing

9.1 Data preprocessing

9.1.1 Data handling

Download. The observables as well as the navigation message and additional information are generally stored in a binary (and receiver dependent) format. The downloading of the data from the receiver is necessary before processing can begin.

Most GPS manufacturers have designed a data management system which they recommend for processing data. Individual manufacturers' software is fully documented in their manuals and will not be covered here. In the following sections a general universal processing scheme is provided which describes the basic principles used.

Data management. As a rule, a single frequency receiver with a 1-second sampling rate collects 0.15 Mb data per satellite during one hour tracking. Hence, one can conclude that during a multisession and multibaseline GPS survey a large amount of data in the Gigabyte range may be collected. In order to archive and to process these data in a reasonable time, an appropriate data structure has to be used. Among different concepts, one based on site/session tables is described here.

Assume a GPS survey including six points and using four receivers. With two overlapping points, two sessions are needed according to Eq. (7.4). In the corresponding site/session table, cf. Table 9.1, the occupation of a site

Table 9.1. Site/session table

Site	Session a	Session b
P1	*	*
P2	*	
P3		*
P4		*
P5	*	
P6	*	*

in a session is marked by an asterisk. This table immediately shows during which sessions a certain site has been occupied, or, inversely, which sites have been included in a certain session. The latter question is more relevant, as it determines which baselines can be computed from a certain session. For this reason, computer storage based on sessions is preferable.

The header of each session file should contain the session identifier and a list of the occupied sites. The header is followed by data blocks. The first block could contain the information for all satellites tracked during that session. An additional block could be reserved to store the following data for each site:

1. Measured data (e.g., carrier phases, code ranges, meteorological data).

2. Intermediate results (e.g., navigation solution, diagnostic messages).

3. Supplementary information (e.g., site description, receiver unit, field crew).

Figure 9.1, for example, represents the file for the first session of the survey. This type of storage is called a linear list after Knuth (1978) where the data are addressed by the use of pointers. To find the data for a certain site, one has simply to search in the header for the site identifier, and the corresponding data are accessed by the pointer.

Data exchange. Although the binary receiver data may have been converted into computer independent ASCII format during downloading, the data are still receiver dependent. In this case, the data management previously de-

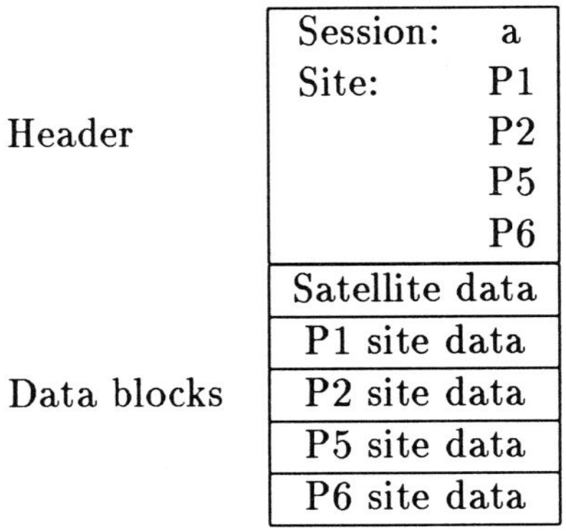

	Session: a
	Site: P1
Header	P2
	P5
	P6
	Satellite data
	P1 site data
Data blocks	P2 site data
	P5 site data
	P6 site data

Fig. 9.1. Data management by a linear list

Table 9.2. Contents of RINEX format

Observation data file	Meteorological data file	Navigation message file
<u>Header</u> Site Crew Equipment Eccentricities Observation types Comments	<u>Header</u> Site Observation types Comments	<u>Header</u> Comments
<u>Data</u> Epoch Satellites Measurements Flags	<u>Data</u> Epoch Measurements	<u>Data</u> Epoch SV clock parameter SV orbital parameters Ionospheric correction Flags

scribed is appropriate only when (in each session) one receiver type is used. Also, each GPS processing software has its own format which necessitates the conversion of specific data into a software independent format when they are processed with a different type of program.

From the preceding, one can conclude that a receiver independent format of GPS data would promote the data exchange. Such a common format should use standard definitions and should also be flexible enough to meet future requirements. To date, many formats have been proposed by various institutions but, for several reasons, they have not received wide acceptance.

A recent attempt has been made with the RINEX (Receiver Independent Exchange) format, cf. Table 9.2. This format was first defined by Gurtner et al. (1989) and has been published in a second version by Gurtner and Mader (1990). The format consists of three types of ASCII files: (1) the observation data file containing the range data, (2) the meteorological data file, and (3) the navigation message file. The records of the files have variable lengths with a maximum of 80 characters per line. Each file is composed of a header section and a data section. The header section contains generic file information and the data section contains the actual data. The navigation message file is site independent, while the observation and meteorological data files must be created for each site of the session.

At present, RINEX is the most favored format. As a consequence, some receiver manufacturers produce software for the conversion of their receiver dependent format into RINEX. The U.S. National Geodetic Survey (NGS) has acted as a coordinator for these efforts. For a detailed description of the RINEX format, the reader is referred to the cited literature. Table 9.2 gives an overview of the different file formats. For the satellites the abbreviation SV, meaning space vehicle, has been used.

RINEX uses the file naming convention "ssssdddf.yyt". The first four characters of the sequence are the site identifier (ssss), the next three the day of year (ddd), and the eighth character is the session number (f). The first two file extension characters denote the last two digits of the current year (yy), and the file type (t) is given by the last character. The satellite designation is in the form "snn". The first character (s) is an identifier of the satellite system and the remaining two digits denote the satellite number (e.g., the PRN number). Thus, the RINEX format enables the possible combination of different satellite observations such as GPS and TRANSIT.

9.1.2 Cycle slip detection and repair

Definition of cycle slips. When a receiver is turned on, the fractional beat phase (i.e., the difference between the satellite transmitted carrier and a receiver generated replica signal) is observed and an integer counter is initialized. During tracking the counter is incremented by one (1) whenever the fractional phase changes from 2π to 0. Thus, at a given epoch the observed accumulated phase $\Delta\varphi$, cf. Remondi (1984) or Remondi (1985), is the sum of the fractional phase φ and the integer count n. The initial integer number N of cycles between the satellite and the receiver is unknown. This phase ambiguity N remains constant as long as no loss of the signal lock occurs. In this event, the integer counter is reinitialized which causes a jump in the instantaneous accumulated phase by an integer number of cycles. This jump is called cycle slip which, of course, is restricted to phase measurements.

A graphical representation of a cycle slip is given in Fig. 9.2. When the measured phases are plotted versus time, a fairly smooth curve should be obtained. In the case of a cycle slip, a sudden jump appears in the plotted curve.

Three sources for cycle slips can be distinguished. First, cycle slips are caused by obstructions of the satellite signal due to trees, buildings, bridges, mountains, etc. This source is the most frequent one. The second source for cycle slips is a low signal-to-noise ratio due to bad ionospheric conditions, multipath, high receiver dynamics, or low satellite elevation. A third source is a failure in the receiver software, cf. Hein (1990b), which leads to incorrect signal processing. Cycle slips could also be caused by malfunctioning satellite

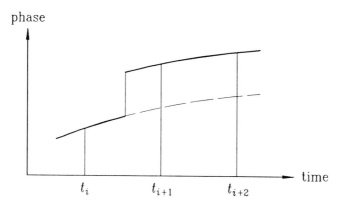

phase

t_i t_{i+1} t_{i+2} time

Fig. 9.2. Graphical representation of a cycle slip

oscillators, but these cases are rare.

As seen from Fig. 9.2, cycle slip detection and repair requires the location of the jump (i.e., cycle slip) and the determination of its size. Repairs are made by correcting all subsequent phase observations for this satellite and this carrier by a fixed amount. Detection is accomplished by a testing quantity. In the example given, this is the measured raw phase. The determination of the cycle slip size and the correction of the phase data is often denoted as cycle slip "fixing".

Testing quantities. The formulation of testing quantities is based on measured carrier phases and code ranges. For a single site, the testing quantities are phases, phase combinations, or combinations of phases and code ranges. Single receiver tests are important because they enable in situ cycle slip detection and repair by the receiver's internal software. When two sites are involved, single-, double-, and triple-differences, cf. Sect. 8.2.1, provide testing quantities. Table 9.3 summarizes a number of candidate testing quantities.

The measured phase $\Phi_i^j(t)$ can be modeled by

$$\lambda \, \Phi_i^j(t) = \varrho_i^j(t) + \lambda \, N_i^j + c \, \Delta\delta_i^j(t) - \frac{A_i^j(t)}{f^2} + \ldots \tag{9.1}$$

where i and j denote the site and the satellite, respectively. The term $A_i^j(t)$ has been substituted for $(40.3\,\mathrm{TEC}/\cos z')$ according to Eq. (6.59). Note that the phase model contains a number of time dependent terms on the right-hand side which may prevent cycle slip detection.

The model for the dual frequency phase combination is developed considering a single site and a single satellite. Thus, the sub- and superscripts

Table 9.3. Testing quantities to detect cycle slips

Required data	Testing quantity	
	Single site	Two sites
Single frequency phase ($L1$ or $L2$)	Undifferenced phase	Single-difference Double-difference Triple-difference
Dual frequency phases ($L1$ and $L2$)	Phase combination (ionospheric residual)	
Single frequency phase ($L1$ or $L2$) and code range	Phase/code range combination	

in Eq. (9.1) can be omitted; whereas, the frequency dependency is shown explicitly by L_1 and L_2:

$$\lambda_{L1}\,\Phi_{L1}(t) = \varrho(t) + \lambda_{L1}\,N_{L1} + c\,\Delta\delta(t) - \frac{A(t)}{f_{L1}^2} + \dots$$

$$\lambda_{L2}\,\Phi_{L2}(t) = \varrho(t) + \lambda_{L2}\,N_{L2} + c\,\Delta\delta(t) - \frac{A(t)}{f_{L2}^2} + \dots \ . \tag{9.2}$$

For the difference of the two equations

$$\lambda_{L1}\,\Phi_{L1}(t) - \lambda_{L2}\,\Phi_{L2}(t) = \lambda_{L1}\,N_{L1} - \lambda_{L2}\,N_{L2} - \frac{A(t)}{f_{L1}^2} + \frac{A(t)}{f_{L2}^2}$$

$$\tag{9.3}$$

the frequency independent terms (i.e., the geometric distance and the clock error) vanish. Dividing (9.3) by λ_{L1} provides

$$\Phi_{L1}(t) - \frac{\lambda_{L2}}{\lambda_{L1}}\,\Phi_{L2}(t) = N_{L1} - \frac{\lambda_{L2}}{\lambda_{L1}}\,N_{L2} - \frac{A(t)}{\lambda_{L1}}\left(\frac{1}{f_{L1}^2} - \frac{1}{f_{L2}^2}\right)$$

$$\tag{9.4}$$

which may be slightly transformed using $c = \lambda\,f$ and from this one obtains

$$\frac{\lambda_{L2}}{\lambda_{L1}} = \frac{f_{L1}}{f_{L2}} \tag{9.5}$$

so that

$$\Phi_{L1}(t) - \frac{f_{L1}}{f_{L2}}\, \Phi_{L2}(t) = N_{L1} - \frac{f_{L1}}{f_{L2}}\, N_{L2} - \frac{A(t)}{\lambda_{L1}}\left(\frac{1}{f_{L1}^2} - \frac{1}{f_{L2}^2}\right)$$

(9.6)

or

$$\Phi_{L1}(t) - \frac{f_{L1}}{f_{L2}}\, \Phi_{L2}(t) = N_{L1} - \frac{f_{L1}}{f_{L2}}\, N_{L2} - \frac{A(t)}{\lambda_{L1}\, f_{L1}^2}\left(1 - \frac{f_{L1}^2}{f_{L2}^2}\right)$$

(9.7)

which is the final form of the dual frequency combination. This model is often denoted as the ionospheric residual, cf. Goad (1986). The right-hand side of Eq. (9.7) shows that the ionospheric residual does not contain time varying terms except for the ionospheric refraction. In comparison to the influence on the raw phases in Eq. (9.1), the influence of the ionosphere on the dual frequency combination is reduced by a factor $(1 - (f_{L1}/f_{L2})^2)$. Substituting the appropriate values for f_{L1} and f_{L2}, yields a reduction of 65%, cf. Goad (1986).

If there are no cycle slips, the temporal variations of the ionospheric residual would be small for normal ionospheric conditions and for short baselines. Indicators of cycle slips are sudden jumps in successive values of the ionospheric residual. The remaining problem is to determine if the cycle slip was on $L1$, $L2$, or both. This will be investigated in the next paragraph.

Note that the ionospheric residual

$$\Phi_{L1}(t) - \frac{f_{L1}}{f_{L2}}\, \Phi_{L2}(t)$$

is a scaled difference of dual frequency phases just as the ionospheric-free linear combination

$$\Phi_{L1}(t) - \frac{f_{L2}}{f_{L1}}\, \Phi_{L2}(t)\,,$$

cf. Eq. (6.79). These two expressions differ, essentially, by the reciprocal nature of the Φ_{L2} coefficients.

Another testing quantity follows from a phase/code range combination. Modeling the carrier phase and the code pseudoranges by

$$\lambda\, \Phi_i^j(t) = \varrho_i^j(t) + \lambda\, N_i^j + c\, \Delta\delta_i^j(t) - \Delta^{Iono}(t) + \Delta^{Trop}$$

$$R_i^j(t) \;= \varrho_i^j(t) \qquad\quad + c\, \Delta\delta_i^j(t) + \Delta^{Iono}(t) + \Delta^{Trop}$$

(9.8)

and forming the difference

$$\lambda \, \Phi_i^j(t) - R_i^j(t) = \lambda \, N_i^j - 2 \, \Delta^{Iono}(t) \tag{9.9}$$

provides a formula where the time dependent terms (except the ionospheric refraction) vanish from the right-hand side. Thus, the phase/code range combination could also be used as testing quantity. The ionospheric influence may either be modeled, cf. for example Beutler et al. (1987), or neglected. One might justify neglecting the ionospheric term since the change of $\Delta^{Iono}(t)$ will be fairly small between closely spaced epochs.

The simple testing quantity (9.9) has a shortcoming which is related to the noise level. The noise level is in the range of ten cycles for time series of the phase/code range combinations, cf. Bastos and Landau (1988). This noise is mainly caused by the noise level of the code measurements and to a minor extent by the ionosphere. The noise of code measurements is larger than the noise for phase measurements because resolution and multipath are proportional to the wavelength. Traditionally, the measurement resolution was $\lambda/100$; today's receiver hardware are achieving improved measurement resolutions approaching $\lambda/1000$. In other words, this would lead to (P-) code range noise levels of a few centimeters. Hence, the phase/code range combination would be an ideal testing quantity for cycle slip detection.

Many authors use single-, double- or triple-differences for cycle slip detection, cf. for example Goad (1985), Remondi (1985), Hilla (1986), Beutler et al. (1987). This means that, in a first step, unrepaired phase combinations are used to process an approximate baseline vector. The corresponding residuals are then tested. Quite often several iterations are necessary to improve the baseline solution. Note that triple-differences can achieve convergence and rather high accuracy without fixing cycle slips.

The list of testing quantities in Table 9.3 is not complete. Allison and Eschenbach (1989) use differences of integrated Doppler and Bock and Shimada (1989) propose the wide lane signal, cf. Eq. (6.16), as testing quantity.

Detection and repair. Each of the described testing quantities allows the location of cycle slips by checking the difference of two consecutive epoch values. This also yields an approximate size of the cycle slip. To find the correct size, the time series of the testing quantity must be investigated in more detail. Note that in case of phases, phase/code range combinations, single-, double-, and triple-differences the detected cycle slip must be an integer. This is not true for the ionospheric residual.

One of the methods for cycle slip detection is the scheme of differences. The principle can be seen from an example given in Lichtenegger and Hofmann-Wellenhof (1989). Assume $y(t_i)$, $i = 1, 2, \ldots, 7$ as a time series for a signal which contains a jump of ε at epoch t_4:

t_i	$y(t)$	y^1	y^2	y^3	y^4
t_1	0				
		0			
t_2	0		0		
		0		ε	
t_3	0		ε		-3ε
		ε		-2ε	
t_4	ε		$-\varepsilon$		3ε
		0		ε	
t_5	ε		0		$-\varepsilon$
		0		0	
t_6	ε		0		
		0			
t_7	ε				

In this scheme, y^1, y^2, y^3, y^4 denote the first-order, second-order, third-order and fourth-order differences. The important property in the context of data irregularities is the amplification of a jump in higher-order differences and thus the improved possibility of detecting the jump. The theoretical reason implied is the fact that differences are generated by subtractive filters. These are high-pass filters damping low frequencies and eliminating constant parts. High frequency constituents such as a jump are amplified. Replacing the signal $y(t)$ for example by the phase and assuming ε to be a cycle slip, the effect of the scheme of differences becomes evident. Any of the quantities of Table 9.3 may be used as signal $y(t)$ for the scheme of differences.

A method to determine the size of a cycle slip is to fit a curve through the testing quantities before and after the cycle slip. The size of the cycle slip is found from the shift between the two curves. The fits may be obtained from a simple linear regression, cf. Mader (1986), or from more realistic least squares models, cf. Beutler et al. (1984) or Cross and Ahmad (1988). These methods are generally called interpolation techniques. Other possibilities are prediction methods such as the Kalman filtering. At a certain epoch the function value (i.e., one of the testing quantities) for the next epoch is predicted based on the information obtained from preceding function values. The predicted value is then compared with the observed value to detect a cycle slip. The application of Kalman filtering for cycle slip detection is proposed by e.g. Goad (1986) and demonstrated in detail by e.g. Landau (1988). More details on the Kalman filtering are provided in Sect. 9.2.2.

When a cycle slip has been detected (by one of the methods previously discussed), the testing quantities can be corrected by adding the size of the cycle slip to each of the subsequent quantities. Hilla (1986) and Cross and Ahmad (1988) report the details of this process.

The assignment of the detected cycle slip to a single phase observation is ambiguous in the case the testing quantities were phase combinations. An exception is the ionospheric residual. Under special circumstances, this testing quantity permits a unique separation, cf. Lichtenegger and Hofmann-Wellenhof (1989). Consider Eq. (9.7) and assume ambiguity changes ΔN_{L1} and ΔN_{L2} caused by cycle slips. Consequently, a jump ΔN in the ionospheric residual would be detected. This jump is equivalent to

$$\Delta N = \Delta N_{L1} - \frac{f_{L1}}{f_{L2}} \Delta N_{L2} . \tag{9.10}$$

Note that ΔN is not an integer. Equation (9.10) represents a diophantine equation for the two integer unknowns ΔN_{L1} and ΔN_{L2}. One equation and two unknowns; hence, there is no unique solution. This can be seen by solving for integer values ΔN_{L1} and ΔN_{L2} such that ΔN becomes zero. To get $\Delta N = 0$, the condition

$$\frac{\Delta N_{L1}}{\Delta N_{L2}} = \frac{f_{L1}}{f_{L2}} = \frac{154}{120} \tag{9.11}$$

must be fulfilled which may be rewritten as

$$\Delta N_{L1} = \frac{77}{60} \Delta N_{L2} . \tag{9.12}$$

As an example, this means that $\Delta N_{L1} = 77$ and $\Delta N_{L2} = 60$ cannot be distinguished from $\Delta N_{L1} = 154$ and $\Delta N_{L2} = 120$ since both solutions satisfy Eq. (9.12). However, the solution would be unambiguous if ΔN_{L1} is less than 77 cycles. The consideration so far assumed error free measurements. To be more realistic, the effect of measurement noise must be taken into account. A simple model for the phase measurement noise is

$$m_\Phi = \pm 0.01 \text{ cycles} \tag{9.13}$$

which corresponds to a resolution of $\lambda/100$. The same model is applied to both carriers $L1$ and $L2$ and thus frequency dependent noise such as multipath is neglected. The assumption is not correct for codeless receivers since additional noise is introduced by the squaring technique, cf. Hofmann-Wellenhof and Lichtenegger (1988), p. 43.

The value ΔN, in principle, is derived from two consecutive ionospheric residuals. Hence,

$$\Delta N(\Delta t) = \Phi_{L1}(t + \Delta t) - \frac{f_{L1}}{f_{L2}} \Phi_{L2}(t + \Delta t)$$
$$- \left[\Phi_{L1}(t) - \frac{f_{L1}}{f_{L2}} \Phi_{L2}(t) \right] \tag{9.14}$$

and applying to this equation the error propagation law gives

$$m_{\Delta N(\Delta t)} = \pm 2.3 \, m_\Phi = \pm 0.023 \, \text{cycles} . \tag{9.15}$$

The 3σ-error therefore yields approximately ± 0.07 cycles. This can be interpreted as the resolution of ΔN. The conclusion is that two ΔN calculated by (9.10) and using arbitrary integers ΔN_{L1} and ΔN_{L2} must differ by at least 0.07 cycles in order to be uniquely separable. A systematic investigation of the lowest values for ΔN_{L1}, ΔN_{L2} is given in Table 9.4. For ΔN_{L1} and ΔN_{L2} the values 0, ± 1, ± 2, ..., ± 5 have been permutated and ΔN calculated by (9.10). Table 9.4 is sorted with increasing ΔN in column one. In the second column the difference of the respective two neighboring function values ΔN is given. To shorten the length of the table, only the negative function values ΔN and zero are displayed. For supplementing with positive function values, the three signs in a row must be reversed.

Those rows in Table 9.4 being marked with an asterisk do not fulfill the criterion of an at least 0.07 cycle difference. For these values an unambiguous separation is not possible because the measurement noise is larger than the separation value. Consider the next to the last row in Table 9.4. A jump in the ionospheric residual of about 0.14 cycle could result from the pair of cycle slips $\Delta N_{L1} = -4$, $\Delta N_{L2} = -3$ or $\Delta N_{L1} = 5$, $\Delta N_{L2} = 4$; however, notice that for the marked lines either ΔN_{L1} or ΔN_{L2} equals 5 (plus or minus). Therefore, omitting the values for $\Delta N_{L1} = \pm 5$ and $\Delta N_{L2} = \pm 5$ creates uniqueness in the sense of separability. Up to ± 4 cycles the function values ΔN are discernible by 0.12 cycles.

The conclusions reached for the cycle slip method based on the dual frequency data combination are as follows. Based on the measurement noise assumption in (9.13), the separation of the cycle slips is unambiguously possible for up to ± 4 cycles. Using a smaller measurement noise increases the separability. When repairing larger cycle slips, another method should be used in order to avoid making a wrong choice in ambiguous situations. Westrop et al. (1989) mention a phase/code range combination but cite possible failures since the code range noise can be larger than ± 4 cycles.

Often, there will be more than one cycle slip. In these cases, each cycle slip must be detected and corrected in turn. The corrected phases, single-, double- or triple-differences are then used to process the baseline.

Table 9.4. Resulting ΔN by permutating ΔN_{L1} and ΔN_{L2}

ΔN	Diff.	ΔN_{L1}	ΔN_{L2}
-11.42	1.00	-5	5
-10.42	0.29	-4	5
-10.13	0.71	-5	4
-9.42	0.29	-3	5
-9.13	0.28	-4	4
-8.85	0.43	-5	3
-8.42	0.29	-2	5
-8.13	0.28	-3	4
-7.85	0.29	-4	3
-7.56	0.14	-5	2
-7.42	0.29	-1	5
-7.13	0.28	-2	4
-6.85	0.29	-3	3
-6.56	0.14	-4	2
-6.42	0.14	0	5
-6.28	0.15	-5	1
-6.13	0.28	-1	4
-5.85	0.29	-2	-3
-5.56	0.14	-3	2
-5.42	0.14	1	5
-5.28	0.15	-4	1
-5.13	0.13	0	4
-5.00	0.15	-5	0
-4.85	0.29	-1	3
-4.56	0.14	-2	2
-4.42	0.14	2	5
-4.28	0.15	-3	1
-4.13	0.13	1	4
-4.00	0.15	-4	0
-3.85	0.13	0	3
-3.72		-5	-1

ΔN	Diff.	ΔN_{L1}	ΔN_{L2}	
-3.72	0.16	-5	-1	
-3.56	0.14	-1	2	
-3.42	0.14	3	5	
-3.28	0.15	-2	1	
-3.13	0.13	2	4	
-3.00	0.15	-3	0	
-2.85	0.13	1	3	
-2.72	0.16	-4	-1	
-2.56	0.12	0	2	
-2.44	0.02	-5	-2	
-2.42	0.14	4	5	*
-2.28	0.15	-1	1	
-2.13	0.13	3	4	
-2.00	0.15	-2	0	
-1.85	0.13	2	3	
-1.52	0.16	-3	-1	
-1.56	0.12	1	2	
-1.44	0.02	-4	-2	
-1.42	0.14	5	5	*
-1.28	0.13	0	1	
-1.15	0.02	-5	-3	*
-1.13	0.13	4	4	
-1.00	0.15	-1	0	
-0.85	0.13	3	3	
-0.72	0.16	-2	-1	
-0.56	0.12	2	2	
-0.44	0.16	-3	-2	
-0.28	0.13	1	1	
-0.15	0.02	-4	-3	
-0.13	0.13	5	4	*
0.00		0	0	

9.1.3   Ambiguity resolution

The ambiguity inherent with phase measurements depends upon both the receiver and the satellite. There is no time dependency as long as tracking is maintained without interruption. In the model for the phase,

$$\Phi = \frac{1}{\lambda}\varrho + f\,\Delta\delta + N - \frac{1}{\lambda}\Delta^{Iono}\,, \tag{9.16}$$

the ambiguity is denoted by N. As soon as the ambiguity is determined as an integer value, the ambiguity is said to be resolved or fixed, cf. Counselman and Abbot (1989). This is very important for the solution of the baseline vectors since, in general, the ambiguity resolution strengthens the baseline solution, cf. Goad (1985). There are, however, some exceptional situations. Rothacher et al. (1989) demonstrate an example where the solutions with fixed ambiguities and floating ambiguities (i.e., real values) agree within a few millimeters. Nevertheless, resolution of the ambiguities generally improves vector accuracy.

Only a few key principles from the numerous kinds of ambiguity resolution techniques will be demonstrated. Many variations may then be derived from these basic methods.

Resolving ambiguities with single frequency phase data. When phase measurements for only one frequency (either $L1$ or, possibly, $L2$) are available, the most direct approach is as follows. The measurements are modeled by Eq. (9.16), and the linearized equations are processed. Depending on the model chosen, a number of unknowns (e.g., point coordinates, clock parameters, etc.) are estimated along with N in a common adjustment. In this geometric approach the unmodeled errors affect all estimated parameters. Therefore, the integer nature of the ambiguities is lost and they are estimated as real values. Many error sources affect the closeness of the estimated ambiguities to their integer values. Some of these sources are: the incomplete phase model, the baseline length (because of varying atmospheric conditions), and orbital errors. To fix ambiguities to integer values, a sequential adjustment could be performed, cf. Blewitt (1987). After an initial adjustment, the ambiguity with a computed value closest to an integer and with minimum standard error is considered to be determined most reliably. This bias is then fixed and the adjustment is repeated (with one less unknown) to fix another ambiguity and so on. When using double-differences over short baselines, this approach is usually successful. The critical factor is the ionospheric refraction which must be modeled and which may prevent a correct resolution of all ambiguities.

Resolving ambiguities with dual frequency phase data. The situation for the ambiguity resolution changes significantly when using dual frequency phase data. There are many advantages implied in dual frequency data because of the various possible linear combinations that can be formed. Wide lane and narrow lane techniques have been proposed e.g. by Counselman et al. (1979), Hatch (1982), Hatch and Larson (1985). Denoting the $L1$ and $L2$ phase data Φ_{L1} and Φ_{L2}, then, according to Eq. (6.16),

$$\Phi_w = \Phi_{L1} - \Phi_{L2} \tag{9.17}$$

is the wide lane signal. The frequency of this signal is $f_w = 347.82\,\text{MHz}$ and the corresponding wavelength $\lambda_w = 86.2\,\text{cm}$. This is a significant increase compared to the original wavelengths of 19.0 and 24.4 cm. The increased wide lane wavelength λ_w provides an increased ambiguity spacing. This is the key to easier resolution of the integer ambiguities. To show the principle, consider the phase models for the carriers $L1$ and $L2$:

$$\Phi_{L1} = \frac{1}{\lambda_{L1}}\varrho + f_{L1}\,\Delta\delta + N_{L1} - \frac{1}{\lambda_{L1}}\Delta^{Iono}(f_{L1})$$

$$\Phi_{L2} = \frac{1}{\lambda_{L2}}\varrho + f_{L2}\,\Delta\delta + N_{L2} - \frac{1}{\lambda_{L2}}\Delta^{Iono}(f_{L2})\,. \tag{9.18}$$

Replacing λ by frequency f via $\lambda = c/f$ and substituting for the ionospheric term $\Delta^{Iono}(f) = A/f^2$ correspondingly, leads to

$$\Phi_{L1} = \frac{f_{L1}}{c}\varrho + f_{L1}\,\Delta\delta + N_{L1} - \frac{A}{c\,f_{L1}}$$

$$\Phi_{L2} = \frac{f_{L2}}{c}\varrho + f_{L2}\,\Delta\delta + N_{L2} - \frac{A}{c\,f_{L2}}\,. \tag{9.19}$$

The difference of the two equations gives

$$\Phi_w = \frac{f_w}{c}\varrho + f_w\,\Delta\delta + N_w - \frac{A}{c}\left(\frac{1}{f_{L1}} - \frac{1}{f_{L2}}\right) \tag{9.20}$$

with the wide lane quantities

$$\Phi_w = \Phi_{L1} - \Phi_{L2}$$

$$f_w = f_{L1} - f_{L2}$$

$$N_w = N_{L1} - N_{L2}\,.$$

The adjustment based on the wide lane model gives wide lane ambiguities N_w which are more easily resolved than the base carrier ambiguities. To compute the ambiguities for the measured phases (e.g., for $L1$), divide the first equation of (9.19) by f_{L1} and (9.20) by f_w, and form the difference of the two equations. This gives

$$\frac{1}{f_{L1}}\Phi_{L1} - \frac{1}{f_w}\Phi_w = \frac{N_{L1}}{f_{L1}} - \frac{N_w}{f_w} - \frac{A}{c\,f_{L1}^2} + \frac{A}{c\,f_w}\left(\frac{1}{f_{L1}} - \frac{1}{f_{L2}}\right) \tag{9.21}$$

and the desired ambiguity N_{L1} follows explicitly after rearranging and multiplying the equation above by f_{L1}:

$$N_{L1} = \Phi_{L1} - (\Phi_w - N_w)\frac{f_{L1}}{f_w} + \frac{A}{c\,f_{L1}} - \frac{A}{c\,f_w}\left(1 - \frac{f_{L1}}{f_{L2}}\right)\,. \tag{9.22}$$

Before commenting on this result, the terms reflecting the ionospheric influence are grouped as

$$+\ \frac{A}{c}\ \frac{f_w\, f_{L2} - f_{L1}\, f_{L2} + f_{L1}^2}{f_{L1}\, f_w\, f_{L2}} \tag{9.23}$$

which is equivalent to

$$+\ \frac{A}{c}\ \frac{f_w\, f_{L2} + f_{L1}\,(f_{L1} - f_{L2})}{f_{L1}\, f_w\, f_{L2}}\ . \tag{9.24}$$

The term in parentheses is replaced by the wide lane frequency which then cancels. Thus,

$$+\ \frac{A}{c}\ \frac{f_{L2} + f_{L1}}{f_{L1}\, f_{L2}} \tag{9.25}$$

results from the ionospheric terms. Therefore, the phase ambiguity N_{L1} in (9.22) can be calculated from the wide lane ambiguity by

$$N_{L1} = \Phi_{L1} - (\Phi_w - N_w)\frac{f_{L1}}{f_w} + \frac{A}{c}\ \frac{f_{L1} + f_{L2}}{f_{L1}\, f_{L2}} \tag{9.26}$$

and in an analogous way for N_{L2} by replacing $L1$ by $L2$ and vice versa in the equation above. Note that the distance ϱ and the clock bias term $\Delta\delta$ have dropped out. These terms are, however, implicitly contained in N_w, cf. Eq. (9.20). The ionospheric term is most annoying. This term will cancel for short baselines with similar ionospheric refraction at both endpoints (using differenced phases). For long baselines or irregular ionospheric conditions, however, the ionospheric term may cause problems.

Linear combinations other than wide laning have been considered such as the ionospheric-free linear combination Φ_{L3}. The disadvantage of this combination is that the corresponding ambiguity is no longer an integer. This is a kind of circulus vitiosus: either the ambiguities may be resolved where the ionosphere is a problem or the ionospheric influence is eliminated which destroys the integer nature of the ambiguities. Wübbena (1989) shows the use of other linear combinations ranging from narrow lane with a 10.7 cm wavelength to extra wide laning with a 172.4 cm wavelength.

Resolving ambiguities by combining dual frequency carrier phase and code data. The most unreliable factor of the wide lane technique described in the previous paragraph is the influence of the ionosphere which increases with baseline length. This drawback can be eliminated by a combination of phase and code data. Consider the models for dual frequency carrier phases and code ranges

$$\Phi_{L1} = \frac{\varrho}{c} f_{L1} - \frac{\kappa}{f_{L1}} + N_{L1} + f_{L1} \Delta\delta \tag{9.27}$$

$$\Phi_{L2} = \frac{\varrho}{c} f_{L2} - \frac{\kappa}{f_{L2}} + N_{L2} + f_{L2} \Delta\delta \tag{9.28}$$

$$R_{L1} = \frac{\varrho}{c} f_{L1} + \frac{\kappa}{f_{L1}} \qquad\quad + f_{L1} \Delta\delta \tag{9.29}$$

$$R_{L2} = \frac{\varrho}{c} f_{L2} + \frac{\kappa}{f_{L2}} \qquad\quad + f_{L2} \Delta\delta \tag{9.30}$$

where both are expressed in cycles of the corresponding carrier. The ionospheric term has been simplified by substituting $\kappa = A/c$. Note that four equations are available with four unknowns for each epoch. The unknowns are $(\varrho/c + \Delta\delta)$, κ, and the ambiguities N_{L1}, N_{L2}.

The stepwise calculation of the ambiguities makes use of the wide lane ambiguity $N_w = N_{L1} - N_{L2}$. The derivation of this ambiguity is lengthy and the impatient reader may continue with (9.42).

By forming the differences between the corresponding carrier and code phases

$$\begin{aligned}
\Phi_{L1} - R_{L1} &= -\frac{2\kappa}{f_{L1}} + N_{L1} \\[2mm]
\Phi_{L2} - R_{L2} &= -\frac{2\kappa}{f_{L2}} + N_{L2}
\end{aligned} \tag{9.31}$$

the geometric distance and the clock bias term are eliminated. The difference of the two equations above yields

$$\Phi_w - R_{L1} + R_{L2} = 2\kappa \left(\frac{1}{f_{L2}} - \frac{1}{f_{L1}} \right) + N_w \tag{9.32}$$

where the wide lane signal Φ_w and the wide lane ambiguity N_w have been substituted. Introducing a common denominator for the expression in the parentheses, gives

$$\Phi_w - R_{L1} + R_{L2} = 2\kappa \frac{f_{L1} - f_{L2}}{f_{L1} f_{L2}} + N_w . \tag{9.33}$$

Now, the ionospheric term κ can be calculated by dividing the code ranges (9.29) by f_{L1} and (9.30) by f_{L2}

$$\begin{aligned}
\frac{R_{L1}}{f_{L1}} &= \frac{\varrho}{c} + \frac{\kappa}{f_{L1}^2} + \Delta\delta \\[2mm]
\frac{R_{L2}}{f_{L2}} &= \frac{\varrho}{c} + \frac{\kappa}{f_{L2}^2} + \Delta\delta
\end{aligned} \tag{9.34}$$

and by subtracting the two equations from each other

$$\frac{R_{L1}}{f_{L1}} - \frac{R_{L2}}{f_{L2}} = \kappa \left(\frac{1}{f_{L1}^2} - \frac{1}{f_{L2}^2} \right). \tag{9.35}$$

This may be written

$$\frac{f_{L2} R_{L1} - f_{L1} R_{L2}}{f_{L1} f_{L2}} = \kappa \frac{f_{L2}^2 - f_{L1}^2}{f_{L1}^2 f_{L2}^2} \tag{9.36}$$

or

$$\frac{f_{L2} R_{L1} - f_{L1} R_{L2}}{f_{L1} f_{L2}} = \kappa \frac{(f_{L2} - f_{L1})}{f_{L1} f_{L2}} \frac{(f_{L2} + f_{L1})}{f_{L1} f_{L2}} \tag{9.37}$$

where the denominator cancels. Finally, rewriting the equation and dividing by $f_{L1} + f_{L2}$ yields

$$\frac{f_{L2} R_{L1} - f_{L1} R_{L2}}{f_{L1} + f_{L2}} = -\kappa \frac{f_{L1} - f_{L2}}{f_{L1} f_{L2}} \tag{9.38}$$

which may now be substituted into (9.33) to eliminate κ. This step in detail gives

$$\Phi_w - R_{L1} + R_{L2} = -2 \frac{f_{L2} R_{L1} - f_{L1} R_{L2}}{f_{L1} + f_{L2}} + N_w \tag{9.39}$$

which may be rearranged to

$$\begin{aligned} N_w = \Phi_w + \frac{1}{f_{L1} + f_{L2}} &\, [(-R_{L1} + R_{L2})(f_{L1} + f_{L2}) \\ &+ 2 f_{L2} R_{L1} - 2 f_{L1} R_{L2}] \end{aligned} \tag{9.40}$$

or

$$\begin{aligned} N_w = \Phi_w + \frac{1}{f_{L1} + f_{L2}} &\, [-R_{L1} f_{L1} + R_{L2} f_{L2} \\ &+ f_{L2} R_{L1} - f_{L1} R_{L2}] \end{aligned} \tag{9.41}$$

and finally

$$N_w = \Phi_w - \frac{f_{L1} - f_{L2}}{f_{L1} + f_{L2}} (R_{L1} + R_{L2}). \tag{9.42}$$

This rather elegant equation allows for the determination of the wide lane ambiguity N_w for each epoch and each site. It is independent of the baseline length and of the ionospheric effects. Even if all modeled systematic effects cancel out in (9.42), the multipath effect remains and affects phase and code differently, cf. Sect. 6.5. Multipath is almost exclusively responsible for a variation of N_w by several cycles from epoch to epoch, cf. Dong and Bock (1989). These variations may be overcome by averaging over a longer period.

Having calculated the wide lane ambiguities, the use of (9.26) would lead to the ambiguities of the undifferenced phases which are effected by the ionospheric refraction. This can be avoided by the following procedure. Start again with the phase equations

$$
\Phi_{L1} = \frac{f_{L1}}{c} \varrho + f_{L1} \Delta\delta + N_{L1} - \frac{\kappa}{f_{L1}}
$$
$$
\Phi_{L2} = \frac{f_{L2}}{c} \varrho + f_{L2} \Delta\delta + N_{L2} - \frac{\kappa}{f_{L2}}
$$
(9.43)

and multiply the first equation by f_{L1} and the second by f_{L2}. The resulting equations are then differenced and thus,

$$
f_{L2}\,\Phi_{L2} - f_{L1}\,\Phi_{L1} = \frac{\varrho}{c}\,(f_{L2}^2 - f_{L1}^2) + \Delta\delta\,(f_{L2}^2 - f_{L1}^2)
$$
$$
+ f_{L2}\,N_{L2} - f_{L1}\,N_{L1}
$$
(9.44)

is obtained. Introducing the wide lane ambiguity by $N_{L2} = N_{L1} - N_w$ leads to

$$
f_{L2}\,\Phi_{L2} - f_{L1}\,\Phi_{L1} = \frac{\varrho}{c}\,(f_{L2}^2 - f_{L1}^2) + \Delta\delta\,(f_{L2}^2 - f_{L1}^2)
$$
$$
- f_{L2}\,N_w + N_{L1}\,(f_{L2} - f_{L1})
$$
(9.45)

or, with $f_w = f_{L1} - f_{L2}$,

$$
N_{L1} = \frac{f_{L1}}{f_w}\,\Phi_{L1} - \frac{f_{L2}}{f_w}\,(\Phi_{L2} + N_w) - (\frac{\varrho}{c} + \Delta\delta)\,(f_{L1} + f_{L2}) \quad (9.46)
$$

which is an ionospherically free equation for the determination of the N_{L1} ambiguities. Note that N_w has been determined by (9.42). A final remark concerning the formula above is appropriate. Combining the terms containing N_w and N_{L1} into a single term (for the ionospherically free model), destroys the terms' integer nature. This integer nature can be preserved by separately calculating the ambiguities, first N_w using (9.42) and then N_{L1} by (9.46).

The approach described is not the only one. Hatch (1982), Bender and Larden (1985), Melbourne (1985), Wübbena (1985), Hatch (1986) show other procedures combining carrier phase and code data. This technique is also appropriate for instantaneous ambiguity resolution in kinematic applications, cf. Hatch (1990).

If there is at least one redundant (fifth) satellite available, least squares techniques can be applied. Loomis (1989) uses, based on a Kalman filter technique, the redundant satellite information in the state vector. The number of states is matched to the number of satellites and varies from 8–12 for 4–8 satellites in view. Three states account for the position, one for the receiver clock, and the remaining for the integer ambiguities. Hwang (1990) shows a variation of the Loomis (1989) approach.

When comparing various methods, consideration of the necessary epochs to be measured is important. Hatch (1990) mentions that a single epoch solution is usually possible if (for short baselines) seven or more satellites can be tracked. The latter requirement is the most severe restriction of this method. Following Loomis (1989), a very long convergence time for the correct ambiguity solution is necessary. On the other hand, this technique does not rely on the P-code. The other methods fall somewhere in between with regard to the necessary epochs.

Search techniques. To aid the search technique, the covariance information resulting from the adjusted (real value) ambiguities may be used. Within an appropriate region centered about the real value solution, all integer values are considered as candidates for the ambiguity. Then all possible combinations of these ambiguities are considered as known values and substituted into subsequent adjustments. From the set of solutions, the integer combination giving the smallest root mean square (rms) error for the position is taken as the "best" solution. This solution is the one chosen when the ratio of the rms of fit between the "best" and "next best" solution is greater than two to three. This method is only practical for a single baseline.

In a previous paragraph another search technique was mentioned, namely the sequential fixing of the ambiguities. This method is possible but time consuming. A slightly more sophisticated method is proposed by Frei and Beutler (1989). This method is also based on the statistical information resulting from a preliminary (baseline) adjustment. The technique can be summarized as follows: (1) estimate the ambiguities and the position to be determined by a standard adjustment, (2) use the statistical information determined by the adjustment to fix the ambiguities one at a time, (3) the resolved ambiguities are substituted as known values and the procedure is repeated. Finally (4), all ambiguities are known integer values and the data are readjusted to determine a definitive solution, cf. Frei (1991).

An alternate technique substitutes the position as known and solves for the ambiguities as unknowns. This could be performed in the following way.

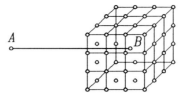

Fig. 9.3. Search technique

Eliminate the ambiguities by forming triple-differences and obtain a first estimate for the position and its standard deviation σ by an adjustment. Center now the approximate position within a cube of dimension $8\sigma \times 8\sigma \times 8\sigma$ (i.e., $\pm 4\sigma$ in each direction and for each coordinate) and partition the cube into a regular grid. The cube thus contains a matrix of points where the center point is the triple-difference solution, see Fig. 9.3. Each of these grid points is considered as candidate for the correct solution. Consequently, one by one, each candidate position is substituted into the observation equation. Then the adjustment (holding the trial position fixed) is performed and the ambiguities are computed. When all points within the cube have been considered, select the solution where the estimated real values of the ambiguities appear as close as possible to integer values. Now, fix the ambiguities to these integer values and compute (holding the ambiguities fixed) the final position which will, in general, be slightly different from the corresponding grid point of the cube.

Ambiguity function method. The same principle of the gridded cube is used for the ambiguity function method. Counselman and Gourevitch (1981) proposed the principle of the ambiguity function and Remondi (1984), Remondi (1990a), Mader (1990) further investigated the concept. The idea should become clear from the following description. Assume a model for the single-difference phase represented by

$$\Phi^j_{AB}(t) = \frac{2\pi}{\lambda} \left[\varrho^j_{AB}(t) + \lambda\, N^j_{AB} - c\, \delta_{AB}(t) \right] \tag{9.47}$$

for the points A and B, and the satellite j, where the dimension cycle has been replaced by radians. Assuming point A as known and B as a candidate from the gridded cube, the known term may be shifted to the left-hand side of the equation:

$$\Phi^j_{AB}(t) - \frac{2\pi}{\lambda} \varrho^j_{AB}(t) = 2\pi\, N^j_{AB} - 2\pi\, f\, \delta_{AB}(t). \tag{9.48}$$

The key is to circumvent the ambiguities N^j_{AB}. A special effect occurs if the term $2\pi\, N^j_{AB}$ is used as the argument of a cosine- or sine-function because

N_{AB}^j is an integer. Therefore, the whole expression (9.48) is placed into the complex plane by raising both the left- and right-hand sides to the power of e^i where $i = \sqrt{-1}$ is the imaginary unit. In detail

$$e^{i\{\Phi_{AB}^j(t) - \frac{2\pi}{\lambda} \varrho_{AB}^j(t)\}} = e^{i\{2\pi N_{AB}^j - 2\pi f \, \delta_{AB}(t)\}} \tag{9.49}$$

which may be written as

$$e^{i\{\Phi_{AB}^j(t) - \frac{2\pi}{\lambda} \varrho_{AB}^j(t)\}} = e^{i\, 2\pi N_{AB}^j} \cdot e^{-i\, 2\pi f \, \delta_{AB}(t)} . \tag{9.50}$$

It is illustrative to consider this situation in the complex plane, cf. Fig. 9.4. Note the equivalence

$$e^{i\alpha} = \cos\alpha + i \sin\alpha \tag{9.51}$$

which may be represented as a unit vector with the components $\cos\alpha$ and $\sin\alpha$ if a real axis and an imaginary axis are used. Therefore,

$$e^{i\, 2\pi N_{AB}^j} = \cos(2\pi \, N_{AB}^j) + i \, \sin(2\pi \, N_{AB}^j) = 1 + i \cdot 0 \tag{9.52}$$

results because of the integer nature of N_{AB}^j. Hence, for one epoch and one satellite

$$e^{i\{\Phi_{AB}^j(t) - \frac{2\pi}{\lambda} \varrho_{AB}^j(t)\}} = e^{-i\, 2\pi f \, \delta_{AB}(t)} \tag{9.53}$$

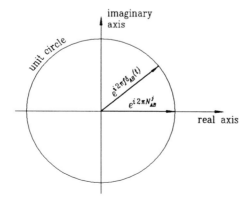

Fig. 9.4. Vector representation in the complex plane

remains. Considering n_j satellites and forming the sum over these satellites for the epoch t leads to

$$\sum_{j=1}^{n_j} e^{i\{\Phi^j_{AB}(t) - \frac{2\pi}{\lambda} \varrho^j_{AB}(t)\}} = n_j \, e^{-i\, 2\pi f\, \delta_{AB}(t)} \,. \tag{9.54}$$

Considering more than one epoch, one must account for the fact that the clock error $\delta_{AB}(t)$ varies with time. A glance at Fig. 9.4 should recall that $e^{-i\, 2\pi f\, \delta_{AB}(t)}$ is a unit vector. Thus, when $\|e^{-i\, 2\pi f\, \delta_{AB}(t)}\| = 1$ is applied to (9.54) one gets

$$\left\| \sum_{j=1}^{n_j} e^{i\{\Phi^j_{AB}(t) - \frac{2\pi}{\lambda} \varrho^j_{AB}(t)\}} \right\| = n_j \cdot 1 \tag{9.55}$$

where the clock error has now vanished.

Take for example four satellites and an error free situation (i.e., neither measurement errors nor model errors, and correct coordinates for the points A and B). In this case the evaluation of the left-hand side of (9.55) should yield 4 where $\Phi^j_{AB}(t)$ are the single-differences of measured phases and $\varrho^j_{AB}(t)$ can be calculated from the known points and satellite positions. However, if point B was chosen incorrectly then the result must be less than 4. In reality, this maximum can probably never be achieved precisely because of measurement errors and incomplete modeling. Thus, the task is restricted to obtain the maximum of (9.55) by varying B.

With highly stable receiver clocks and close epoch spacing it is theoretically possible to include more than one epoch within the absolute value. Using n_t epochs, the contribution of all epochs may be summed up by

$$\sum_{t=1}^{n_t} \left\| \sum_{j=1}^{n_j} e^{i\{\Phi^j_{AB}(t) - \frac{2\pi}{\lambda} \varrho^j_{AB}(t)\}} \right\| = n_t \, n_j \tag{9.56}$$

where for simplicity the same number of satellites at all epochs is assumed. Following Remondi (1984) and Remondi (1990a), the left-hand side (i.e., the double sum) is denoted as an ambiguity function. Analogous to the case with one epoch, the maximum of the ambiguity function must be found. In general it will, as before, be less than the theoretical value $n_t \, n_j$.

The ambiguity function procedure is simple. Assume an approximate solution for point B, e.g. by triple-differences. Then, place this solution into the center of a cube, cf. Fig. 9.3, and partition the cube into grid points. Each grid point is a candidate for the final solution and the ambiguity function (9.56) is calculated for all single-differences. The grid point yielding the maximum ambiguity function value, which should theoretically be equal to the total number of single-differences (i.e., $n_t \, n_j$), is the desired solution. Having found this solution, the ambiguities could be computed using double-differences. Also, an adjustment using double-differences might be performed to verify the position of B and the ambiguities. The computation of point B with fixed ambiguities is the final step.

It is worth noting that the ambiguity function method is completely insensitive to cycle slips. The reason can easily be seen from Eq. (9.52). Even if the ambiguity changes by an arbitrary integer amount ΔN_{AB}^j, then $e^{i\,2\pi\,(N_{AB}^j + \Delta N_{AB}^j)}$ is still unity and the subsequent equations remain therefore unchanged. Other methods require that cycle slips be repaired before computing the ambiguity.

Remondi (1984) shows detailed examples of how to speed up the procedure, how to choose the density of the grid points within the cube, and how to find the correct maximum if there are many relative maxima for the ambiguity function. These considerations are significant, since the computational burden could, otherwise, become overwhelming. For illustrative purposes, assume a $6\,\text{m} \times 6\,\text{m} \times 6\,\text{m}$ cube with a one centimeter grid. Then $(601)^3 \approx 2.17 \cdot 10^8$ possible solutions must be checked with the ambiguity function (9.56).

9.2 Adjustment, filtering, and smoothing

9.2.1 Least squares adjustment

Standard adjustment. There are numerous adjustment techniques that can be used, but least squares adjustment with parameters is the only one discussed here. It is based on equations where the observations are expressed as a function of unknown parameters. A Taylor series expansion is usually performed in the case of nonlinear functions. This requires approximate values for the parameters. The Taylor series expansion must be truncated after the second term to obtain a linear function with respect to the unknowns. The resulting linear observation model can be represented in a matrix-vector notation as

$$\underline{l} = \underline{A}\,\underline{x} \tag{9.57}$$

where

$$
\begin{array}{lll}
\underline{l} & \ldots & \text{vector of observations} \\
\underline{A} & \ldots & \text{design matrix} \\
\underline{x} & \ldots & \text{vector of unknowns.}
\end{array}
$$

By introducing in addition the definitions

$$\sigma_0^2 \quad \dots \quad \text{a priori variance}$$

$$\underline{\Sigma} \quad \dots \quad \text{covariance matrix},$$

the cofactor matrix of observations is

$$\underline{Q}_l = \frac{1}{\sigma_0^2}\underline{\Sigma} \tag{9.58}$$

and

$$\underline{P} = \underline{Q}_l^{-1} \tag{9.59}$$

is the weight matrix. Assuming n observations and u unknown parameters leads to a design matrix $\underline{A}$ comprising n rows and u columns. For $n > u$ the system (9.57) is overdetermined and, in general, nonconsistent because of observational errors or noise. To assure consistency, the noise vector $\underline{n}$ is added to the vector of observations and Eq. (9.57) thus converts to

$$\underline{l} + \underline{n} = \underline{A}\,\underline{x}. \tag{9.60}$$

The solution of this system becomes unique by the least squares principle $\underline{n}^T\underline{P}\,\underline{n} = minimum$. The application of this minimum principle on the observation equations (9.60) leads to the normal equations

$$\underline{A}^T\underline{P}\,\underline{A}\,\underline{x} = \underline{A}^T\underline{P}\,\underline{l} \tag{9.61}$$

with the solution

$$\underline{x} = (\underline{A}^T\underline{P}\,\underline{A})^{-1}\underline{A}^T\underline{P}\,\underline{l} \tag{9.62}$$

which can be simplified to

$$\underline{x} = \underline{G}^{-1}\underline{g} \tag{9.63}$$

where $\underline{G} = \underline{A}^T\underline{P}\,\underline{A}$ and $\underline{g} = \underline{A}^T\underline{P}\,\underline{l}$.

The cofactor matrix $\underline{Q}_x$ follows from $\underline{x} = \underline{G}^{-1}\underline{A}^T\underline{P}\,\underline{l}$ by the covariance propagation law as

$$\underline{Q}_x = (\underline{G}^{-1}\underline{A}^T\underline{P})\underline{Q}_l(\underline{G}^{-1}\underline{A}^T\underline{P})^T \tag{9.64}$$

and reduces to

$$\underline{Q}_x = \underline{G}^{-1} = (\underline{A}^T\underline{P}\,\underline{A})^{-1} \tag{9.65}$$

by substituting $\underline{Q}_l = \underline{P}^{-1}$.

Sequential adjustment. Assume a partitioning of the observation model (9.60) into two subsets:

$$l = \begin{bmatrix} l_1 \\ l_2 \end{bmatrix} \qquad n = \begin{bmatrix} n_1 \\ n_2 \end{bmatrix} \qquad A = \begin{bmatrix} A_1 \\ A_2 \end{bmatrix}. \tag{9.66}$$

Using the first set only, a preliminary solution $x_{(0)}$ can be calculated according to (9.62) and (9.65) by

$$\begin{aligned} x_{(0)} &= (A_1^T P_1 A_1)^{-1} A_1^T P_1 l_1 = G_1^{-1} g_1 \\ Q_{x_{(0)}} &= (A_1^T P_1 A_1)^{-1} = G_1^{-1}. \end{aligned} \tag{9.67}$$

Provided that there is no correlation between the two subsets of observations, the weight matrix

$$P = \begin{bmatrix} P_1 & 0 \\ 0 & P_2 \end{bmatrix} \tag{9.68}$$

is a block-diagonal matrix. The matrix G and the vector g for the adjustment of the full set of observations result from adding the corresponding matrices and vectors for the two subsets:

$$\begin{aligned} G &= A^T P A = (A_1^T P_1 A_1 + A_2^T P_2 A_2) = G_1 + G_2 \\ g &= A^T P l = (A_1^T P_1 l_1 + A_2^T P_2 l_2) = g_1 + g_2. \end{aligned} \tag{9.69}$$

If the change of the preliminary solution $x_{(0)}$ due to the additional observation set l_2 is denoted as Δx, then

$$(G_1 + G_2)(x_{(0)} + \Delta x) = g_1 + g_2 \tag{9.70}$$

is the appropriate formulation of the adjustment. This equation can be slightly rearranged to

$$(G_1 + G_2)\Delta x = g_1 + g_2 - (G_1 + G_2)x_{(0)} \tag{9.71}$$

where the right-hand side, cf. Eq. (9.67), can be simplified because of the relation $g_1 - G_1 x_{(0)} = 0$ so that

$$(G_1 + G_2)\Delta x = g_2 - G_2 x_{(0)} \tag{9.72}$$

results. Resubstituting from (9.69) $g_2 = A_2^T P_2 l_2$ and $G_2 = A_2^T P_2 A_2$ yields

$$(G_1 + G_2)\Delta x = A_2^T P_2 l_2 - A_2^T P_2 A_2 x_{(0)} \tag{9.73}$$

or

$$(\underline{G}_1 + \underline{G}_2)\,\Delta\underline{x} = \underline{A}_2^T \underline{P}_2\,(\underline{l}_2 - \underline{A}_2\,\underline{x}_{(0)}) \tag{9.74}$$

and

$$\Delta\underline{x} = (\underline{G}_1 + \underline{G}_2)^{-1}\underline{A}_2^T \underline{P}_2\,(\underline{l}_2 - \underline{A}_2\,\underline{x}_{(0)}) \tag{9.75}$$

or finally

$$\Delta\underline{x} = \underline{K}\,(\underline{l}_2 - \underline{A}_2\,\underline{x}_{(0)}) \tag{9.76}$$

where

$$\underline{K} = (\underline{G}_1 + \underline{G}_2)^{-1}\underline{A}_2^T \underline{P}_2\,. \tag{9.77}$$

Note that formally in (9.76) the term $\underline{A}_2\,\underline{x}_{(0)}$ can be considered as prediction for the observations $\underline{l}_2$.

The goal of the next step is the computation of the change $\Delta\underline{Q}$ with respect to the preliminary cofactor matrix $\underline{Q}_{x_{(0)}}$. Starting point is the relation

$$\underline{G}\,\underline{Q}_x = (\underline{G}_1 + \underline{G}_2)(\underline{Q}_{x_{(0)}} + \Delta\underline{Q}) = \underline{I} \tag{9.78}$$

where $\underline{I}$ denotes the unit matrix. This equation is reformulated as

$$(\underline{G}_1 + \underline{G}_2)\,\Delta\underline{Q} = \underline{I} - (\underline{G}_1 + \underline{G}_2)\,\underline{Q}_{x_{(0)}} \tag{9.79}$$

and, since $\underline{G}_1\,\underline{Q}_{x_{(0)}} = \underline{I}$, this reduces to

$$(\underline{G}_1 + \underline{G}_2)\,\Delta\underline{Q} = -\underline{G}_2\,\underline{Q}_{x_{(0)}} \tag{9.80}$$

or

$$\Delta\underline{Q} = -(\underline{G}_1 + \underline{G}_2)^{-1}\underline{G}_2\,\underline{Q}_{x_{(0)}} \tag{9.81}$$

and, by using $\underline{G}_2 = \underline{A}_2^T \underline{P}_2\,\underline{A}_2$, the relation

$$\Delta\underline{Q} = -(\underline{G}_1 + \underline{G}_2)^{-1}\underline{A}_2^T \underline{P}_2\,\underline{A}_2\,\underline{Q}_{x_{(0)}} \tag{9.82}$$

follows. Comparing this equation with (9.77), $\underline{K}$ may be substituted and

$$\Delta\underline{Q} = -\underline{K}\,\underline{A}_2\,\underline{Q}_{x_{(0)}} \tag{9.83}$$

results. Matrix $\underline{K}$, which is denoted as gain matrix, satisfies the very remarkable relation

$$\underline{K} = (\underline{G}_1 + \underline{G}_2)^{-1} A_2^T P_2 = \underline{G}_1^{-1} A_2^T (\underline{P}_2^{-1} + A_2 \underline{G}_1^{-1} A_2^T)^{-1} . \qquad (9.84)$$

This relation is based on a formula found by Bennet (1965). For additional information cf. also Moritz (1980), Eqs. (19-12) and (19-13). The point of this equation is its application to the inversion of modified matrices of the type $(\underline{C} + \underline{D})$ where $\underline{C}^{-1}$ is known a priori. The identity of (9.84) may be proved by multiplying both sides from left with $(\underline{G}_1 + \underline{G}_2)$ and from right with $(\underline{P}_2^{-1} + A_2 \underline{G}_1^{-1} A_2^T)$.

It is essential to learn from Eq. (9.84) that the first form for $\underline{K}$ implies the inversion of a $u \times u$ matrix if u is the number of unknown parameters; whereas, for the second form an inversion of an $n_2 \times n_2$ matrix is necessary when n_2 denotes the number of observations for the second subset. Therefore, the second form is advantageous as long as $n_2 < u$.

A final remark should conclude the section on the sequential adjustment. In the equation of $\Delta \underline{x}$, cf. (9.76), (9.77), and in the equation for $\Delta \underline{Q}$, cf. (9.83), (9.77), neither the design matrix $\underline{A}_1$ nor the vector $\underline{l}_1$ for the first set of observations appears explicitly. Therefore, formally, the substitution e.g. $\underline{A}_1 = \underline{I}$ and $\underline{l}_1 = \underline{x}_{(0)}$ may be performed and the model for the sequential adjustment is then formulated as

$$\begin{aligned} \underline{x}_{(0)} + \underline{n}_1 &= \underline{x} \\ \underline{l}_2 + \underline{n}_2 &= A_2 \underline{x} . \end{aligned} \qquad (9.85)$$

Model (9.85) reflects that the preliminary estimates $\underline{x}_{(0)}$ for the unknown parameters are introduced into the sequential adjustment as observations. This approach is often used in the context with Kalman filtering, cf. de Munck (1989).

9.2.2 Kalman filtering

Introduction. Consider a dynamic system such as a moving vehicle. The unknown parameters, e.g. the coordinates and the velocity, form the elements of the "state vector". This time dependent vector may be predicted for any instant t by means of "system equations". The predicted values can be improved or updated by observations containing information on some components of the state vector.

The whole procedure is known as Kalman filtering. It corresponds to the sequential adjustment in the static case, cf. Egge (1985). Consequently, optimal estimations of the unknowns on the basis of all observations up to the epoch t are obtained. Note, however, there is no need to store these data for the subsequent epochs.

Prediction. The time dependent state vector $\underline{x}(t)$ comprising the unknown parameters of the dynamic system may be modeled by a system of differential equations of the first order as

$$\underline{\dot{x}}(t) = \underline{F}(t)\,\underline{x}(t) + \underline{w}(t) \tag{9.86}$$

where

$$\begin{array}{lll} \underline{\dot{x}}(t) & \ldots & \text{time derivative of the state vector} \\ \underline{F}(t) & \ldots & \text{dynamics matrix} \\ \underline{w}(t) & \ldots & \text{driving noise.} \end{array}$$

For the following, at the initial epoch t_0 the state vector $\underline{x}(t_0)$ and its cofactor matrix Q_{x_0} are assumed to be known. A general solution for the system equation (9.86) only exists if the matrix $\underline{F}(t)$ contains periodic or constant coefficients. For the latter case this solution can be written as

$$\begin{aligned} \underline{x}(t) &= \underline{T}(t,t_0)\,\underline{x}(t_0) + \int\limits_{t_0}^{t} \underline{T}(t,\tau)\,\underline{w}(\tau)\,d\tau \\ &= \underline{T}(t,t_0)\,\underline{x}(t_0) + \underline{e}(t)\,, \end{aligned} \tag{9.87}$$

cf. Schwarz (1983), Eq. (47). Here, $\underline{T}$ represents the transition matrix. This matrix results from a Taylor series expansion of the state vector. Thus,

$$\underline{x}(t) = \underline{x}(t_0) + \underline{\dot{x}}(t_0)\,(t - t_0) + \frac{1}{2}\,\underline{\ddot{x}}(t_0)\,(t - t_0)^2 + \ldots \tag{9.88}$$

which can be written by substituting (9.86) and neglecting the driving noise $\underline{w}(t)$ as

$$\underline{x}(t) = \underline{x}(t_0) + \underline{F}(t_0)\,\underline{x}(t_0)\,(t - t_0) + \frac{1}{2}\,\underline{F}(t_0)^2\,\underline{x}(t_0)\,(t - t_0)^2 + \ldots \tag{9.89}$$

and, by comparing with (9.87), as

$$\underline{x}(t) = \underline{T}(t,t_0)\,\underline{x}(t_0)\,. \tag{9.90}$$

Consequently, by introducing the substitution $\Delta t = t - t_0$, the transition matrix can be written as an infinite series with respect to $\underline{F}$ by

$$\begin{aligned} \underline{T}(t,t_0) &= \underline{I} + \underline{F}(t_0)\,\Delta t + \frac{1}{2}\,\underline{F}(t_0)^2\,\Delta t^2 + \ldots \\ &= \sum_{n=0}^{\infty} \frac{1}{n!}\,\underline{F}(t_0)^n\,\Delta t^n\,. \end{aligned} \tag{9.91}$$

The cofactor matrix $\underline{Q}_x$ of the state vector $\underline{x}(t)$ can be calculated via (9.87) by using the law of covariance propagation yielding

$$\underline{Q}_x = \underline{T}(t,t_0)\,\underline{Q}_{x_0}\,\underline{T}^T(t,t_0) + \underline{Q}_e \,. \tag{9.92}$$

Studying Eqs. (9.87) and (9.91), it can be concluded that the essential problems of the Kalman filtering are the definition of the transition matrix $\underline{T}$ and of the cofactor matrix $\underline{Q}_e$.

Update. Starting at the initial epoch t_0, the state vector $\underline{x}(t)$ can be predicted for any arbitrary future epoch t by the system equations (9.87). Taking into account the system noise $\underline{e}(t)$, it is assumed that at epoch t observations $\underline{l}(t)$ and the corresponding cofactors $\underline{Q}_l$ are available. These data may be related to the updated state vector $\hat{\underline{x}}(t)$ – possibly after a necessary linearization – by the equation

$$\underline{l}(t) = \underline{A}\,\hat{\underline{x}}(t)\,. \tag{9.93}$$

Considering both the predicted state vector $\underline{x}(t)$ and the observations $\underline{l}(t)$ as stochastic quantities leads to the sequential adjustment problem

$$\begin{aligned} \underline{x}(t) + \underline{n}_x &= \hat{\underline{x}}(t) \\ \underline{l}(t) + \underline{n}_l &= \underline{A}\,\hat{\underline{x}}(t)\,. \end{aligned} \tag{9.94}$$

This system is equivalent to (9.85). Therefore, the solution can be taken immediately from (9.76) and (9.83) by matching the notations to the present situation and it follows

$$\begin{aligned} \hat{\underline{x}}(t) &= \underline{x}(t) + \Delta\underline{x}(t) = \underline{x}(t) + \underline{K}\,[\underline{l}(t) - \underline{A}\,\underline{x}(t)] \\ \hat{\underline{Q}}_x &= \underline{Q}_x + \Delta\underline{Q}_x = (\underline{I} - \underline{K}\,\underline{A})\,\underline{Q}_x\,. \end{aligned} \tag{9.95}$$

The gain matrix $\underline{K}$ is now given as

$$\underline{K} = \underline{Q}_x\,\underline{A}^T(\underline{Q}_l + \underline{A}\,\underline{Q}_x\,\underline{A}^T)^{-1}\,, \tag{9.96}$$

cf. also Eq. (9.84).

Example. Consider a vehicle moving on a straight line with constant velocity v where the motion is affected by the random acceleration a. Also, assume that the (one-dimensional) position $p(t_0)$ and the velocity $v(t_0)$ as well as the corresponding variances σ_p^2, σ_v^2 and that of the noise, σ_a^2, are known at the initial epoch t_0. Furthermore, it is also assumed that the position of the vehicle is observed at an epoch $t = t_0 + \Delta t$ where the observation has the variance σ_l^2.

The state vector consists of the position and the velocity of the vehicle. For the initial epoch, one thus obtains

$$\underline{x}(t_0) = \begin{bmatrix} p(t_0) \\ v(t_0) \end{bmatrix} \quad \text{and} \quad \underline{\dot{x}}(t_0) = \begin{bmatrix} \dot{p}(t_0) \\ \dot{v}(t_0) \end{bmatrix} = \begin{bmatrix} v(t_0) \\ 0 \end{bmatrix}. \tag{9.97}$$

The substitution of these vectors and of the random acceleration a into (9.86) yields the dynamics matrix and the driving noise vector for epoch t_0:

$$\underline{F}(t_0) = \begin{bmatrix} 0 & 1 \\ 0 & 0 \end{bmatrix} \quad \underline{w}(t_0) = a \begin{bmatrix} 0 \\ 1 \end{bmatrix}. \tag{9.98}$$

The transition matrix thus is, cf. Eq. (9.91),

$$\underline{T}(t, t_0) = \begin{bmatrix} 1 & \Delta t \\ 0 & 1 \end{bmatrix}. \tag{9.99}$$

Assuming a constant acceleration during the integration interval Δt, then from (9.87) the noise vector is obtained as

$$\underline{e}(t) = a \begin{bmatrix} \frac{\Delta t^2}{2} \\ \Delta t \end{bmatrix}. \tag{9.100}$$

Note that under the present assumptions the elements of the predicted state vector $\underline{x}(t)$ would also result from the formulas of accelerated motion.

The cofactor matrix $\underline{Q}_x$ of the predicted state vector $\underline{x}(t)$ follows from Eq. (9.92) as

$$\underline{Q}_x = \begin{bmatrix} \sigma_p^2 + \Delta t^2\, \sigma_v^2 + \frac{1}{4}\,\Delta t^4\, \sigma_a^2 & \Delta t\, \sigma_v^2 + \frac{1}{2}\,\Delta t^3\, \sigma_a^2 \\ \Delta t\, \sigma_v^2 + \frac{1}{2}\,\Delta t^3\, \sigma_a^2 & \sigma_v^2 + \Delta t^2\, \sigma_a^2 \end{bmatrix}$$

$$= \begin{bmatrix} Q_{11} & Q_{12} \\ Q_{12} & Q_{22} \end{bmatrix}. \tag{9.101}$$

Since the observation equation is $l(t) + n(t) = \hat{p}(t)$, the matrix $\underline{A}$ in (9.94) shrinks to the row vector

$$\underline{A} = \begin{bmatrix} 1 & 0 \end{bmatrix}, \tag{9.102}$$

and the gain matrix reduces to the column vector

$$\underline{K} = \frac{1}{Q_{11} + \sigma_1^2} \begin{bmatrix} Q_{11} \\ Q_{12} \end{bmatrix}. \tag{9.103}$$

Now, the updated state vector $\underline{\hat{x}}(t)$ and the corresponding cofactor matrix $\underline{\hat{Q}}_x$ can be calculated by Eqs. (9.95).

9.2.3 Smoothing

The process of improving previous estimates for the state vector by a new measurement is called smoothing. Since smoothing is performed backwards in time, it is (contrary to the real-time Kalman filtering) a postmission process. One smoothing technique is presented here for the sake of completeness. For further details the reader is referred to the literature, e.g. Schwarz and Arden (1985). Using the notations

$$\underline{x}(t) \ldots \text{predicted state vector}$$

$$\underline{\hat{x}}(t) \ldots \text{updated state vector}$$

$$\underline{\overset{\circ}{x}}(t) \ldots \text{smoothed state vector}$$

then, following Schwarz (1983), Eqs. (61), (63), (64), a formula for optimal smoothing is

$$\underline{\overset{\circ}{x}}(t_i) = \underline{\hat{x}}(t_i) + \underline{D}(t_{i+1}, t_i) \left[\underline{\overset{\circ}{x}}(t_{i+1}) - \underline{x}(t_{i+1}) \right] \tag{9.104}$$

where the gain matrix is

$$\underline{D}(t_{i+1}, t_i) = \underline{\hat{Q}}_{x_i} \, \underline{T}(t_{i+1}, t_i) \, \underline{Q}_{x_{i+1}}^{-1} . \tag{9.105}$$

At the epoch of the last update measurement, the updated state vector is set identical to the smoothed one and the backwards algorithm can be started. From Eq. (9.104) it may be concluded that the process requires the predicted and updated vectors and the cofactor matrices at the update epochs as well as the transition matrices between the updates. This implies, in general, a large amount of data. This is probably the reason why optimal smoothing is very often replaced by empirical methods.

9.3 Adjustment of mathematical GPS models

9.3.1 Linearization

When the models of Chap. 8 are considered, the only term comprising unknowns in nonlinear form is ϱ. This section explains in detail how is ϱ linearized. The basic formula from Eq. (8.2)

$$\begin{aligned} \varrho_i^j(t) &= \sqrt{(X^j(t) - X_i)^2 + (Y^j(t) - Y_i)^2 + (Z^j(t) - Z_i)^2} \\ &\equiv f(X_i, Y_i, Z_i) \end{aligned} \tag{9.106}$$

shows the unknown point coordinates X_i, Y_i, Z_i in nonlinear form. Assuming certain approximate values X_{i0}, Y_{i0}, Z_{i0} for the unknowns, an approximate distance $\varrho_{i0}^j(t)$ can be calculated accordingly:

$$\varrho_{i0}^j(t) = \sqrt{(X^j(t) - X_{i0})^2 + (Y^j(t) - Y_{i0})^2 + (Z^j(t) - Z_{i0})^2}$$
$$\equiv f(X_{i0}, Y_{i0}, Z_{i0}). \tag{9.107}$$

Having available those values, the unknowns X_i, Y_i, Z_i can be decomposed by

$$X_i = X_{i0} + \Delta X_i$$
$$Y_i = Y_{i0} + \Delta Y_i \tag{9.108}$$
$$Z_i = Z_{i0} + \Delta Z_i$$

where now ΔX_i, ΔY_i, ΔZ_i are new unknowns. This means that the original unknowns have been split into a known part (represented by the approximate values X_{i0}, Y_{i0}, Z_{i0}) and an unknown part (represented by ΔX_i, ΔY_i, ΔZ_i). The advantage of this splitting-up is that the function $f(X_i, Y_i, Z_i)$ is replaced by an equivalent function $f(X_{i0} + \Delta X_i, Y_{i0} + \Delta Y_i, Z_{i0} + \Delta Z_i)$ which can now be expanded by a Taylor series with respect to the approximate point. This leads to

$$f(X_i, Y_i, Z_i) \equiv f(X_{i0} + \Delta X_i, Y_{i0} + \Delta Y_i, Z_{i0} + \Delta Z_i) =$$
$$f(X_{i0}, Y_{i0}, Z_{i0}) + \frac{\partial f(X_{i0}, Y_{i0}, Z_{i0})}{\partial X_{i0}} \Delta X_i \tag{9.109}$$
$$+ \frac{\partial f(X_{i0}, Y_{i0}, Z_{i0})}{\partial Y_{i0}} \Delta Y_i + \frac{\partial f(X_{i0}, Y_{i0}, Z_{i0})}{\partial Z_{i0}} \Delta Z_i + \ldots$$

where the expansion is truncated after the linear term; otherwise, the unknowns ΔX_i, ΔY_i, ΔZ_i would appear in nonlinear form. The partials are obtained from (9.107) by

$$\frac{\partial f(X_{i0}, Y_{i0}, Z_{i0})}{\partial X_{i0}} = -\frac{X^j(t) - X_{i0}}{\varrho_{i0}^j(t)}$$
$$\frac{\partial f(X_{i0}, Y_{i0}, Z_{i0})}{\partial Y_{i0}} = -\frac{Y^j(t) - Y_{i0}}{\varrho_{i0}^j(t)} \tag{9.110}$$
$$\frac{\partial f(X_{i0}, Y_{i0}, Z_{i0})}{\partial Z_{i0}} = -\frac{Z^j(t) - Z_{i0}}{\varrho_{i0}^j(t)}$$

and are the components of the unit vector pointing from the satellite towards the approximate site. The substitution of Eqs. (9.107) and (9.110) into Eq. (9.109) gives

$$\varrho_i^j(t) = \varrho_{i0}^j(t) - \frac{X^j(t) - X_{i0}}{\varrho_{i0}^j(t)} \Delta X_i - \frac{Y^j(t) - Y_{i0}}{\varrho_{i0}^j(t)} \Delta Y_i$$
$$-\frac{Z^j(t) - Z_{i0}}{\varrho_{i0}^j(t)} \Delta Z_i \tag{9.111}$$

where the equivalence of $f(X_i, Y_i, Z_i)$ with $\varrho_i^j(t)$ has been used for the left-hand side. The equation above is now linear with respect to the unknowns ΔX_i, ΔY_i, ΔZ_i.

9.3.2 Linear model for point positioning with code ranges

The model is given only in its elementary form and, thus, apart from the geometry only the clocks are modeled and the ionosphere, troposphere, and other minor effects are neglected. According to Eq. (8.5) the model for point positioning with code ranges is given by

$$R_i^j(t) = \varrho_i^j(t) + c\,\delta^j(t) - c\,\delta_i(t) \tag{9.112}$$

which can be linearized by substituting (9.111):

$$R_i^j(t) = \varrho_{i0}^j(t) - \frac{X^j(t) - X_{i0}}{\varrho_{i0}^j(t)} \Delta X_i - \frac{Y^j(t) - Y_{i0}}{\varrho_{i0}^j(t)} \Delta Y_i$$
$$-\frac{Z^j(t) - Z_{i0}}{\varrho_{i0}^j(t)} \Delta Z_i + c\,\delta^j(t) - c\,\delta_i(t)\,. \tag{9.113}$$

Leaving the terms containing unknowns on the right-hand side, the equation above is rewritten as

$$R_i^j(t) - \varrho_{i0}^j(t) - c\,\delta^j(t) = -\frac{X^j(t) - X_{i0}}{\varrho_{i0}^j(t)} \Delta X_i$$
$$-\frac{Y^j(t) - Y_{i0}}{\varrho_{i0}^j(t)} \Delta Y_i - \frac{Z^j(t) - Z_{i0}}{\varrho_{i0}^j(t)} \Delta Z_i - c\,\delta_i(t) \tag{9.114}$$

where the satellite clock bias is assumed to be known. This assumption makes sense because satellite clock correctors can be received from the navigation message, cf. Sect. 5.1.2. Model (9.114) comprises (for the epoch t)

four unknowns, namely ΔX_i, ΔY_i, ΔZ_i, $\delta_i(t)$. Consequently, four satellites are needed to solve the problem. The shorthand notations

$$l^j = R_i^j(t) - \varrho_{i0}^j(t) - c\,\delta^j(t)$$

$$a_{Xi}^j = -\frac{X^j(t) - X_{i0}}{\varrho_{i0}^j(t)}$$

$$a_{Yi}^j = -\frac{Y^j(t) - Y_{i0}}{\varrho_{i0}^j(t)} \qquad (9.115)$$

$$a_{Zi}^j = -\frac{Z^j(t) - Z_{i0}}{\varrho_{i0}^j(t)}$$

help to simplify the representation of the system of equations. Assuming now four satellites numbered from 1 to 4, then

$$l^1 = a_{Xi}^1\,\Delta X_i + a_{Yi}^1\,\Delta Y_i + a_{Zi}^1\,\Delta Z_i - c\,\delta_i(t)$$

$$l^2 = a_{Xi}^2\,\Delta X_i + a_{Yi}^2\,\Delta Y_i + a_{Zi}^2\,\Delta Z_i - c\,\delta_i(t)$$

$$l^3 = a_{Xi}^3\,\Delta X_i + a_{Yi}^3\,\Delta Y_i + a_{Zi}^3\,\Delta Z_i - c\,\delta_i(t) \qquad (9.116)$$

$$l^4 = a_{Xi}^4\,\Delta X_i + a_{Yi}^4\,\Delta Y_i + a_{Zi}^4\,\Delta Z_i - c\,\delta_i(t)$$

is the appropriate system of equations. Note that the superscripts are the satellite numbers and not exponents! Introducing

$$\underline{A} = \begin{bmatrix} a_{Xi}^1 & a_{Yi}^1 & a_{Zi}^1 & -c \\ a_{Xi}^2 & a_{Yi}^2 & a_{Zi}^2 & -c \\ a_{Xi}^3 & a_{Yi}^3 & a_{Zi}^3 & -c \\ a_{Xi}^4 & a_{Yi}^4 & a_{Zi}^4 & -c \end{bmatrix} \qquad \underline{x}^T = \begin{bmatrix} \Delta X_i \\ \Delta Y_i \\ \Delta Z_i \\ \delta_i(t) \end{bmatrix} \qquad \underline{l} = \begin{bmatrix} l^1 \\ l^2 \\ l^3 \\ l^4 \end{bmatrix}$$

$$(9.117)$$

the set of linear equations can be written in the matrix-vector form

$$\underline{l} = \underline{A}\,\underline{x}. \qquad (9.118)$$

For this first example of a linearized GPS model, the resubstitution for the vector $\underline{l}$ and the matrix $\underline{A}$ is given explicitly for one epoch t:

$$
\underline{l} = \begin{bmatrix} R_i^1(t) - \varrho_{i0}^1(t) - c\,\delta^1(t) \\ R_i^2(t) - \varrho_{i0}^2(t) - c\,\delta^2(t) \\ R_i^3(t) - \varrho_{i0}^3(t) - c\,\delta^3(t) \\ R_i^4(t) - \varrho_{i0}^4(t) - c\,\delta^4(t) \end{bmatrix}
$$

$$
\underline{A} = \begin{bmatrix} -\dfrac{X^1(t) - X_{i0}}{\varrho_{i0}^1(t)} & -\dfrac{Y^1(t) - Y_{i0}}{\varrho_{i0}^1(t)} & -\dfrac{Z^1(t) - Z_{i0}}{\varrho_{i0}^1(t)} & -c \\[2ex] -\dfrac{X^2(t) - X_{i0}}{\varrho_{i0}^2(t)} & -\dfrac{Y^2(t) - Y_{i0}}{\varrho_{i0}^2(t)} & -\dfrac{Z^2(t) - Z_{i0}}{\varrho_{i0}^2(t)} & -c \\[2ex] -\dfrac{X^3(t) - X_{i0}}{\varrho_{i0}^3(t)} & -\dfrac{Y^3(t) - Y_{i0}}{\varrho_{i0}^3(t)} & -\dfrac{Z^3(t) - Z_{i0}}{\varrho_{i0}^3(t)} & -c \\[2ex] -\dfrac{X^4(t) - X_{i0}}{\varrho_{i0}^4(t)} & -\dfrac{Y^4(t) - Y_{i0}}{\varrho_{i0}^4(t)} & -\dfrac{Z^4(t) - Z_{i0}}{\varrho_{i0}^4(t)} & -c \end{bmatrix}.
$$

$$(9.119)$$

No further comment is given on the solution of the linear system (9.118). The coordinate difference ΔX_i, ΔY_i, ΔZ_i and the receiver clock error $\delta_i(t)$ for epoch t result. The desired point coordinates are finally obtained by (9.108).

Point positioning with code ranges is applicable for each epoch separately. Therefore, this adjustment model can also be used in kinematic applications.

9.3.3 Linear model for point positioning with carrier phases

The procedure is the same as in the previous section. In Eq. (8.8) the linearization is performed for $\varrho_i^j(t)$ and known terms are shifted to the left-hand side. The result is

$$
\begin{aligned}
\lambda\,\Phi_i^j(t) - \varrho_{i0}^j(t) - c\,\delta^j(t) &= -\frac{X^j(t) - X_{i0}}{\varrho_{i0}^j(t)}\,\Delta X_i \\
-\frac{Y^j(t) - Y_{i0}}{\varrho_{i0}^j(t)}\,\Delta Y_i &- \frac{Z^j(t) - Z_{i0}}{\varrho_{i0}^j(t)}\,\Delta Z_i + \lambda\,N_i^j - c\,\delta_i(t)
\end{aligned}
$$

$$(9.120)$$

where the number of unknowns is now increased by the ambiguities. Considering again the classical case of four satellites, the system follows as

$$
\underline{l} = \begin{bmatrix} \lambda\,\Phi_i^1(t) - \varrho_{i0}^1(t) - c\,\delta^1(t) \\ \lambda\,\Phi_i^2(t) - \varrho_{i0}^2(t) - c\,\delta^2(t) \\ \lambda\,\Phi_i^3(t) - \varrho_{i0}^3(t) - c\,\delta^3(t) \\ \lambda\,\Phi_i^4(t) - \varrho_{i0}^4(t) - c\,\delta^4(t) \end{bmatrix},
$$

$$
\underline{A} = \begin{bmatrix} -\dfrac{X^1(t)-X_{i0}}{\varrho_{i0}^1(t)} & -\dfrac{Y^1(t)-Y_{i0}}{\varrho_{i0}^1(t)} & -\dfrac{Z^1(t)-Z_{i0}}{\varrho_{i0}^1(t)} & \lambda & 0 & 0 & 0 & -c \\[2mm] -\dfrac{X^2(t)-X_{i0}}{\varrho_{i0}^2(t)} & -\dfrac{Y^2(t)-Y_{i0}}{\varrho_{i0}^2(t)} & -\dfrac{Z^2(t)-Z_{i0}}{\varrho_{i0}^2(t)} & 0 & \lambda & 0 & 0 & -c \\[2mm] -\dfrac{X^3(t)-X_{i0}}{\varrho_{i0}^3(t)} & -\dfrac{Y^3(t)-Y_{i0}}{\varrho_{i0}^3(t)} & -\dfrac{Z^3(t)-Z_{i0}}{\varrho_{i0}^3(t)} & 0 & 0 & \lambda & 0 & -c \\[2mm] -\dfrac{X^4(t)-X_{i0}}{\varrho_{i0}^4(t)} & -\dfrac{Y^4(t)-Y_{i0}}{\varrho_{i0}^4(t)} & -\dfrac{Z^4(t)-Z_{i0}}{\varrho_{i0}^4(t)} & 0 & 0 & 0 & \lambda & -c \end{bmatrix}
$$

$$
\underline{x}^T = \begin{bmatrix} \Delta X_i & \Delta Y_i & \Delta Z_i & N_i^1 & N_i^2 & N_i^3 & N_i^4 & \delta_i(t) \end{bmatrix}
$$

$$(9.121)$$

where the matrix-vector notation has been used. Obviously, the four equations do not solve for the eight unknowns. This reflects the fact that point positioning with phases in this form cannot be solved epoch by epoch. Each additional epoch increases the number of unknowns by a new clock term. Thus, for two epochs there are eight equations and nine unknowns (still an underdetermined problem). For three epochs there are 12 equations and 10 unknowns, thus a slightly overdetermined problem. The 10 unknowns in the latter example are the coordinate increments ΔX_i, ΔY_i, ΔZ_i for the unknown point, the integer ambiguities N_i^1, N_i^2, N_i^3, N_i^4 for the four satellites, and the receiver clock biases $\delta_i(t_1)$, $\delta_i(t_2)$, $\delta_i(t_3)$ for the three epochs.

With the shorthand notations for the coefficients of the coordinate increments

$$
a_{Xi}^j(t_k) = -\frac{X^j(t_k) - X_{i0}}{\varrho_{i0}^j(t_k)}
$$

$$
a_{Yi}^j(t_k) = -\frac{Y^j(t_k) - Y_{i0}}{\varrho_{i0}^j(t_k)}
$$

$$(9.122)$$

$$
a_{Zi}^j(t_k) = -\frac{Z^j(t_k) - Z_{i0}}{\varrho_{i0}^j(t_k)},
$$

the matrix-vector scheme for the example above is given by

$$\underline{l} = \begin{bmatrix} \lambda\,\Phi_i^1(t_1) - \varrho_{i0}^1(t_1) - c\,\delta^1(t_1) \\ \lambda\,\Phi_i^2(t_1) - \varrho_{i0}^2(t_1) - c\,\delta^2(t_1) \\ \lambda\,\Phi_i^3(t_1) - \varrho_{i0}^3(t_1) - c\,\delta^3(t_1) \\ \lambda\,\Phi_i^4(t_1) - \varrho_{i0}^4(t_1) - c\,\delta^4(t_1) \\ \lambda\,\Phi_i^1(t_2) - \varrho_{i0}^1(t_2) - c\,\delta^1(t_2) \\ \lambda\,\Phi_i^2(t_2) - \varrho_{i0}^2(t_2) - c\,\delta^2(t_2) \\ \lambda\,\Phi_i^3(t_2) - \varrho_{i0}^3(t_2) - c\,\delta^3(t_2) \\ \lambda\,\Phi_i^4(t_2) - \varrho_{i0}^4(t_2) - c\,\delta^4(t_2) \\ \lambda\,\Phi_i^1(t_3) - \varrho_{i0}^1(t_3) - c\,\delta^1(t_3) \\ \lambda\,\Phi_i^2(t_3) - \varrho_{i0}^2(t_3) - c\,\delta^2(t_3) \\ \lambda\,\Phi_i^3(t_3) - \varrho_{i0}^3(t_3) - c\,\delta^3(t_3) \\ \lambda\,\Phi_i^4(t_3) - \varrho_{i0}^4(t_3) - c\,\delta^4(t_3) \end{bmatrix} \qquad \underline{x}^T = \begin{bmatrix} \Delta X_i \\ \Delta Y_i \\ \Delta Z_i \\ N_i^1 \\ N_i^2 \\ N_i^3 \\ N_i^4 \\ \delta_i(t_1) \\ \delta_i(t_2) \\ \delta_i(t_3) \end{bmatrix}$$

$$\underline{A} = \begin{bmatrix} a_{Xi}^1(t_1) & a_{Yi}^1(t_1) & a_{Zi}^1(t_1) & \lambda & 0 & 0 & 0 & -c & 0 & 0 \\ a_{Xi}^2(t_1) & a_{Yi}^2(t_1) & a_{Zi}^2(t_1) & 0 & \lambda & 0 & 0 & -c & 0 & 0 \\ a_{Xi}^3(t_1) & a_{Yi}^3(t_1) & a_{Zi}^3(t_1) & 0 & 0 & \lambda & 0 & -c & 0 & 0 \\ a_{Xi}^4(t_1) & a_{Yi}^4(t_1) & a_{Zi}^4(t_1) & 0 & 0 & 0 & \lambda & -c & 0 & 0 \\ a_{Xi}^1(t_2) & a_{Yi}^1(t_2) & a_{Zi}^1(t_2) & \lambda & 0 & 0 & 0 & 0 & -c & 0 \\ a_{Xi}^2(t_2) & a_{Yi}^2(t_2) & a_{Zi}^2(t_2) & 0 & \lambda & 0 & 0 & 0 & -c & 0 \\ a_{Xi}^3(t_2) & a_{Yi}^3(t_2) & a_{Zi}^3(t_2) & 0 & 0 & \lambda & 0 & 0 & -c & 0 \\ a_{Xi}^4(t_2) & a_{Yi}^4(t_2) & a_{Zi}^4(t_2) & 0 & 0 & 0 & \lambda & 0 & -c & 0 \\ a_{Xi}^1(t_3) & a_{Yi}^1(t_3) & a_{Zi}^1(t_3) & \lambda & 0 & 0 & 0 & 0 & 0 & -c \\ a_{Xi}^2(t_3) & a_{Yi}^2(t_3) & a_{Zi}^2(t_3) & 0 & \lambda & 0 & 0 & 0 & 0 & -c \\ a_{Xi}^3(t_3) & a_{Yi}^3(t_3) & a_{Zi}^3(t_3) & 0 & 0 & \lambda & 0 & 0 & 0 & -c \\ a_{Xi}^4(t_3) & a_{Yi}^4(t_3) & a_{Zi}^4(t_3) & 0 & 0 & 0 & \lambda & 0 & 0 & -c \end{bmatrix}$$

$$(9.123)$$

The solution of this overdetermined system is performed by the least squares adjustment procedure as described in Sect. 9.2.

9.3.4   Linear model for relative positioning

The previous sections have shown linear models for both code ranges and carrier phases. For the case of relative positioning, the investigation is restricted to the carrier phases, since it should be obvious how to change from the more expanded model of phases to a code model. Furthermore, the linearization and setup of the system of linear equations remains, in principle, the same for phases and phase combinations and could be performed analogously for each model. Therefore, the double-difference is selected for treatment in detail. The model for the double-difference, cf. Eq. (8.22), is

$$\Phi_{AB}^{jk}(t) = \frac{1}{\lambda}\, \varrho_{AB}^{jk}(t) + N_{AB}^{jk} \tag{9.124}$$

where the term ϱ_{AB}^{jk} containing the geometry is composed as

$$\varrho_{AB}^{jk}(t) = \varrho_B^k(t) - \varrho_B^j(t) - \varrho_A^k(t) + \varrho_A^j(t) \tag{9.125}$$

which reflects the fact of four measurements for a double-difference. Each of the four terms must be linearized according to (9.111) which leads to

$$
\begin{aligned}
\varrho_{AB}^{jk}(t) = {} & \varrho_{B0}^k(t) - \frac{X^k(t) - X_{B0}}{\varrho_{B0}^k(t)}\,\Delta X_B - \frac{Y^k(t) - Y_{B0}}{\varrho_{B0}^k(t)}\,\Delta Y_B \\[2mm]
& \hspace{6.5cm} - \frac{Z^k(t) - Z_{B0}}{\varrho_{B0}^k(t)}\,\Delta Z_B \\[2mm]
& - \varrho_{B0}^j(t) + \frac{X^j(t) - X_{B0}}{\varrho_{B0}^j(t)}\,\Delta X_B + \frac{Y^j(t) - Y_{B0}}{\varrho_{B0}^j(t)}\,\Delta Y_B \\[2mm]
& \hspace{6.5cm} + \frac{Z^j(t) - Z_{B0}}{\varrho_{B0}^j(t)}\,\Delta Z_B \\[2mm]
& - \varrho_{A0}^k(t) + \frac{X^k(t) - X_{A0}}{\varrho_{A0}^k(t)}\,\Delta X_A + \frac{Y^k(t) - Y_{A0}}{\varrho_{A0}^k(t)}\,\Delta Y_A \\[2mm]
& \hspace{6.5cm} + \frac{Z^k(t) - Z_{A0}}{\varrho_{A0}^k(t)}\,\Delta Z_A \\[2mm]
& + \varrho_{A0}^j(t) - \frac{X^j(t) - X_{A0}}{\varrho_{A0}^j(t)}\,\Delta X_A - \frac{Y^j(t) - Y_{A0}}{\varrho_{A0}^j(t)}\,\Delta Y_A \\[2mm]
& \hspace{6.5cm} - \frac{Z^j(t) - Z_{A0}}{\varrho_{A0}^j(t)}\,\Delta Z_A\,.
\end{aligned}
\tag{9.126}
$$

To achieve a linear system in the form $\underline{l} = \underline{A}\,\underline{x}$ it is helpful to introduce the abbreviations

$$a_{XA}^{jk}(t) = +\frac{X^k(t) - X_{A0}}{\lambda\,\varrho_{A0}^k(t)} - \frac{X^j(t) - X_{A0}}{\lambda\,\varrho_{A0}^j(t)}$$

$$a_{YA}^{jk}(t) = +\frac{Y^k(t) - Y_{A0}}{\lambda\,\varrho_{A0}^k(t)} - \frac{Y^j(t) - Y_{A0}}{\lambda\,\varrho_{A0}^j(t)}$$

$$a_{ZA}^{jk}(t) = +\frac{Z^k(t) - Z_{A0}}{\lambda\,\varrho_{A0}^k(t)} - \frac{Z^j(t) - Z_{A0}}{\lambda\,\varrho_{A0}^j(t)}$$

$$a_{XB}^{jk}(t) = -\frac{X^k(t) - X_{B0}}{\lambda\,\varrho_{B0}^k(t)} + \frac{X^j(t) - X_{B0}}{\lambda\,\varrho_{B0}^j(t)} \qquad (9.127)$$

$$a_{YB}^{jk}(t) = -\frac{Y^k(t) - Y_{B0}}{\lambda\,\varrho_{B0}^k(t)} + \frac{Y^j(t) - Y_{B0}}{\lambda\,\varrho_{B0}^j(t)}$$

$$a_{ZB}^{jk}(t) = -\frac{Z^k(t) - Z_{B0}}{\lambda\,\varrho_{B0}^k(t)} + \frac{Z^j(t) - Z_{B0}}{\lambda\,\varrho_{B0}^j(t)}$$

and for the left-hand side

$$l_{AB}^{jk}(t) = \lambda\,\Phi_{AB}^{jk}(t) - \varrho_{B0}^k(t) + \varrho_{B0}^j(t) + \varrho_{A0}^k(t) - \varrho_{A0}^j(t) \qquad (9.128)$$

which comprises both the measurements and all terms computed from the approximate values. Using the abbreviations (9.127) and (9.128), the linear observation equation is given by

$$l_{AB}^{jk}(t) = a_{XA}^{jk}(t)\,\Delta X_A + a_{YA}^{jk}(t)\,\Delta Y_A + a_{ZA}^{jk}(t)\,\Delta Z_A$$

$$+ a_{XB}^{jk}(t)\,\Delta X_B + a_{YB}^{jk}(t)\,\Delta Y_B + a_{ZB}^{jk}(t)\,\Delta Z_B + N_{AB}^{jk}\,\lambda$$

$$(9.129)$$

which reflects the general case with two unknown baseline points. However, the coordinates of one point (e.g., A) must be known for true relative positioning. More specifically, the known point A reduces the number of unknowns by three because of

$$\Delta X_A = \Delta Y_A = \Delta Z_A = 0 \qquad (9.130)$$

and leads to a slight change in the left-hand side term

$$l_{AB}^{jk}(t) = \lambda\,\Phi_{AB}^{jk}(t) - \varrho_{B0}^k(t) + \varrho_{B0}^j(t) + \varrho_{A}^k(t) - \varrho_{A}^j(t). \qquad (9.131)$$

Assuming now four satellites j, k, ℓ, m and two epochs t_1, t_2, the matrix-vector system

$$
\underline{l} =
\begin{bmatrix}
l_{AB}^{jk}(t_1) \\[4pt]
l_{AB}^{j\ell}(t_1) \\[4pt]
l_{AB}^{jm}(t_1) \\[4pt]
l_{AB}^{jk}(t_2) \\[4pt]
l_{AB}^{j\ell}(t_2) \\[4pt]
l_{AB}^{jm}(t_2)
\end{bmatrix}
\qquad
\underline{x} =
\begin{bmatrix}
\Delta X_B \\[4pt]
\Delta Y_B \\[4pt]
\Delta Z_B \\[4pt]
N_{AB}^{jk} \\[4pt]
N_{AB}^{j\ell} \\[4pt]
N_{AB}^{jm}
\end{bmatrix}
$$

$$
\underline{A} =
\begin{bmatrix}
a_{XB}^{jk}(t_1) & a_{YB}^{jk}(t_1) & a_{ZB}^{jk}(t_1) & \lambda & 0 & 0 \\[4pt]
a_{XB}^{j\ell}(t_1) & a_{YB}^{j\ell}(t_1) & a_{ZB}^{j\ell}(t_1) & 0 & \lambda & 0 \\[4pt]
a_{XB}^{jm}(t_1) & a_{YB}^{jm}(t_1) & a_{ZB}^{jm}(t_1) & 0 & 0 & \lambda \\[4pt]
a_{XB}^{jk}(t_2) & a_{YB}^{jk}(t_2) & a_{ZB}^{jk}(t_2) & \lambda & 0 & 0 \\[4pt]
a_{XB}^{j\ell}(t_2) & a_{YB}^{j\ell}(t_2) & a_{ZB}^{j\ell}(t_2) & 0 & \lambda & 0 \\[4pt]
a_{XB}^{jm}(t_2) & a_{YB}^{jm}(t_2) & a_{ZB}^{jm}(t_2) & 0 & 0 & \lambda
\end{bmatrix}
$$

$$(9.132)$$

is obtained which represents a determined and thus solvable system. Note that for one epoch the system has more unknowns than equations.

9.4   Network adjustment

9.4.1   Single baseline solution

The previous sections described the linearization of the observation equations. The adjustment itself (i.e., the solution of the system of linear equations) is a purely computational task to be solved by the computer. It is not the objective of this GPS text to investigate different solution strategies. Extensive details on this subject can be found in Leick (1990). Here, only some remarks are appropriate.

The adjustment principle $\underline{n}^T \underline{P} \, \underline{n} = minimum$ necessitates for the solution the implementation of the weight matrix $\underline{P}$. As shown in Sect. 8.2.2, the phases and the single-differences are uncorrelated whereas double-differences and triple-differences are correlated. The implementation of the double-difference correlation can be accomplished fairly easily or the double-differences can be decorrelated by using a Gram-Schmidt orthonormalization, cf. Remondi (1984). The implementation of the correlation of the triple-differences is more difficult. Furthermore, it is questionable if it is worthwhile to put much effort in a correct triple-difference correlation computation where the noise of the triple-differences always prevents a refined solution.

In the case of an observed network, the use of the single baseline method usually implies a baseline by baseline computation for all possible combinations. Denote by n_i the number of observing sites, then $n_i(n_i-1)/2$ baselines can be calculated. Note that only $n_i - 1$ of them are theoretically independent. The redundant baselines can either be used for misclosure checks or for an additional adjustment of the baseline vectors.

There are other approaches. McArthur et al. (1985) propose three different methods. One of them is the same as the previously mentioned one. The second method restricts the selection to $n_i - 1$ vectors. This means that the optimal baseline configuration must be chosen. The third method is a variation of the first one and encompasses more than one session. All possible baselines are computed for each session. Finally, the resulting vectors of all sessions are subject to a common adjustment.

The disadvantage of the simple single baseline solution from the theoretical point of view is that it is not correct. This is due to the correlation of the simultaneously observed baselines. By solving baseline by baseline, this correlation is ignored. The intercorrelation between baselines is discussed in the next section.

9.4.2 Multipoint solution

In contrast to the baseline by baseline solution, here all points in the network are considered at once. The key difference compared to the single baseline solution is that in the multipoint approach the correlations between the baselines are taken into account.

The principal correlations have been shown in Sect. 8.2.2. The same theoretical aspects also apply to the extended case of a network. The examples are kept as simple as possible to avoid the burden of lengthy formulas.

Single-difference example for a network. Consider three points A, B, C, a single satellite j, and a single epoch t. Thus, two independent baselines can be defined. Taking A as reference site, for the two baselines $A - B$ and $A - C$

the two single-differences

$$\Phi_{AB}^j(t) = \Phi_B^j(t) - \Phi_A^j(t)$$
$$\Phi_{AC}^j(t) = \Phi_C^j(t) - \Phi_A^j(t)$$
(9.133)

can be set up for the one satellite at epoch t. By introducing the vector $\underline{SD}(t)$ for the single-differences, the vector $\underline{\Phi}$ for the phases, and a matrix $\underline{C}$ as

$$\underline{SD} = \begin{bmatrix} \Phi_{AB}^j(t) \\ \Phi_{AC}^j(t) \end{bmatrix} \qquad \underline{C} = \begin{bmatrix} -1 & 1 & 0 \\ -1 & 0 & 1 \end{bmatrix} \qquad \underline{\Phi} = \begin{bmatrix} \Phi_A^j(t) \\ \Phi_B^j(t) \\ \Phi_C^j(t) \end{bmatrix} \quad (9.134)$$

the relation $\underline{SD} = \underline{C}\,\underline{\Phi}$ can be formed. To find the correlation, the covariance propagation law is applied by $\operatorname{cov}(\underline{SD}) = \underline{C}\operatorname{cov}(\underline{\Phi})\,\underline{C}^T$ leading to

$$\operatorname{cov}(\underline{SD}) = \sigma^2\,\underline{C}\,\underline{C}^T \tag{9.135}$$

because of $\operatorname{cov}(\underline{\Phi}) = \sigma^2\,\underline{I}$, cf. Eq. (8.32). Substituting matrix $\underline{C}$ from (9.134) and evaluating the matrix operation yields

$$\operatorname{cov}(\underline{SD}(t)) = \sigma^2 \begin{bmatrix} 2 & 1 \\ 1 & 2 \end{bmatrix} \tag{9.136}$$

which shows, as is to be expected, a correlation of the single-differences of the two baselines. Recall that single-differences of a single baseline are uncorrelated as discovered in Sect. 8.2.2 but single-differences of two baselines with a common point are correlated.

Double-difference example for a network. Since double-differences are already correlated for a single baseline, a correlation must be expected for the network too. Nevertheless, the subsequent slightly larger example will demonstrate the increasing complexity. Assume again three points A, B, C with A as reference site for the two baselines $A - B$ and $A - C$. Consider a single epoch t for four satellites j, k, ℓ, m where j is taken as the reference satellite for the double-differences.

Formulating generally, there are $(n_i - 1)(n_j - 1)$ independent double-differences for n_i points and n_j satellites. For the given example $n_i = 3$ and

$n_j = 4$ and thus 6 double-differences can be written. Based on Eq. (8.25) these are

$$\Phi_{AB}^{jk}(t) = \Phi_B^k(t) - \Phi_B^j(t) - \Phi_A^k(t) + \Phi_A^j(t),$$

$$\Phi_{AB}^{j\ell}(t) = \Phi_B^\ell(t) - \Phi_B^j(t) - \Phi_A^\ell(t) + \Phi_A^j(t),$$

$$\Phi_{AB}^{jm}(t) = \Phi_B^m(t) - \Phi_B^j(t) - \Phi_A^m(t) + \Phi_A^j(t),$$

$$\Phi_{AC}^{jk}(t) = \Phi_C^k(t) - \Phi_C^j(t) - \Phi_A^k(t) + \Phi_A^j(t),$$

$$\Phi_{AC}^{j\ell}(t) = \Phi_C^\ell(t) - \Phi_C^j(t) - \Phi_A^\ell(t) + \Phi_A^j(t),$$

$$\Phi_{AC}^{jm}(t) = \Phi_C^m(t) - \Phi_C^j(t) - \Phi_A^m(t) + \Phi_A^j(t),$$

$$(9.137)$$

for the assumptions made. As in the previous example, a matrix-vector relation is desired. By introducing

$$\underline{C} = \begin{bmatrix} 1 & -1 & 0 & 0 & -1 & 1 & 0 & 0 & 0 & 0 & 0 & 0 \\ 1 & 0 & -1 & 0 & -1 & 0 & 1 & 0 & 0 & 0 & 0 & 0 \\ 1 & 0 & 0 & -1 & -1 & 0 & 0 & 1 & 0 & 0 & 0 & 0 \\ 1 & -1 & 0 & 0 & 0 & 0 & 0 & 0 & -1 & 1 & 0 & 0 \\ 1 & 0 & -1 & 0 & 0 & 0 & 0 & 0 & -1 & 0 & 1 & 0 \\ 1 & 0 & 0 & -1 & 0 & 0 & 0 & 0 & -1 & 0 & 0 & 1 \end{bmatrix}$$

$$\underline{DD} = \begin{bmatrix} \Phi_{AB}^{jk}(t) \\ \Phi_{AB}^{j\ell}(t) \\ \Phi_{AB}^{jm}(t) \\ \Phi_{AC}^{jk}(t) \\ \Phi_{AC}^{j\ell}(t) \\ \Phi_{AC}^{jm}(t) \end{bmatrix} \qquad \underline{\Phi} = \begin{bmatrix} \Phi_A^j(t) \\ \Phi_A^k(t) \\ \Phi_A^\ell(t) \\ \Phi_A^m(t) \\ \Phi_B^j(t) \\ \Phi_B^k(t) \\ \Phi_B^\ell(t) \\ \Phi_B^m(t) \\ \Phi_C^j(t) \\ \Phi_C^k(t) \\ \Phi_C^\ell(t) \\ \Phi_C^m(t) \end{bmatrix}$$

$$(9.138)$$

the relation

$$\underline{DD} = \underline{C}\,\underline{\Phi} \qquad\qquad (9.139)$$

is valid. The covariance follows by

$$\text{cov}(\underline{DD}) = \underline{C}\,\text{cov}(\underline{\Phi})\,\underline{C}^T \tag{9.140}$$

which reduces to

$$\text{cov}(\underline{DD}) = \sigma^2\,\underline{C}\,\underline{C}^T \tag{9.141}$$

because of the uncorrelated phases. Explicitly, the matrix product

$$\underline{C}\,\underline{C}^T = \begin{bmatrix} 4 & 2 & 2 & 2 & 1 & 1 \\ 2 & 4 & 2 & 1 & 2 & 1 \\ 2 & 2 & 4 & 1 & 1 & 2 \\ 2 & 1 & 1 & 4 & 2 & 2 \\ 1 & 2 & 1 & 2 & 4 & 2 \\ 1 & 1 & 2 & 2 & 2 & 4 \end{bmatrix} \tag{9.142}$$

is a full matrix as expected. Finally, the weight matrix

$$\underline{P} = \frac{1}{12} \begin{bmatrix} 6 & -2 & -2 & -3 & 1 & 1 \\ -2 & 6 & -2 & 1 & -3 & 1 \\ -2 & -2 & 6 & 1 & 1 & -3 \\ -3 & 1 & 1 & 6 & -2 & -2 \\ 1 & -3 & 1 & -2 & 6 & -2 \\ 1 & 1 & -3 & -2 & -2 & 6 \end{bmatrix} \tag{9.143}$$

is the inverse of (9.142). Ashkenazi et al. (1987) show, how for a computer implementation the weight matrix can be composed by a Kronecker matrix product. Beutler et al. (1986) and Beutler et al. (1987) give detailed instructions for a computer implementation and also show some results of network campaigns where the correlations have either been totally neglected, introduced in a single-baseline mode, or calculated correctly. For small networks with baselines not exceeding 10 km, the differences of the three methods are in the range of a few millimeters. Clearly, the solution without any correlation deviates from the true values by a greater amount. It is estimated that the single baseline method deviates from the multibaseline (correlated) solution by a maximum of 2σ.

9.4.3   Single baseline versus multipoint solution

The previous paragraphs have shown the difference between the single baseline and the multipoint solution from the theoretical point of view and mentioned a few results from GPS campaigns as shown by Beutler et al. (1987). This section presents some arguments for using one or the other method.

- The correlation is not modeled correctly with the single baseline solution because intercorrelations of baselines are neglected.

- The computer program is, without doubt, much simpler for the single baseline approach.

- The computational time is not a real problem because with a modern software and modern computers the processing of a baseline should only require a few minutes.

- Cycle slips are more easily detected and repaired in the multipoint mode, cf. Beck et al. (1989).

- It takes less effort in the single baseline mode to isolate bad measurements and possibly to eliminate them.

- The economic implementation of the full correlation for a multipoint solution only works properly for networks with the same observation pattern at each receiver site. In the event of numerous data outages it is better to calculate the covariance matrix from the scratch.

- Even in the case of the multipoint approach, it becomes questionable if the correlations can be modeled properly. A very illustrative example is given in Beutler et al. (1990) where single and dual frequency receivers are combined in a network. For the dual frequency receivers, the ionospheric-free combination $L3$ is formed from $L1$ and $L2$ and processed together with the $L1$ data of the single frequency receivers. Thus, a correlation is introduced because of the $L1$ data. A proper modeling of the correlation biases the ionospheric-free $L3$-baseline by the ionosphere of the $L1$-baseline, an effect which is definitely undesirable.

The factors listed should help illustrate the difficulty in deciding which method (i.e., single baseline versus multipoint solution) to use.

9.5   Dilution of Precision

The geometry of the visible satellites is an important factor in achieving high quality results especially for point positioning and kinematic surveying. The geometry changes with time due to the relative motion of the satellites. A measure for the geometry is the Dilution of Precision (DOP) factor.

First, the specific case of four satellites is considered. The linearized observation equations for the point positioning model with pseudoranges are given by Eq. (9.118) and the solution for the (four) unknowns follows from the inverse relation $\underline{x} = \underline{A}^{-1}\,\underline{l}$. The design matrix $\underline{A}$ is given by Eq. (9.119)

$$\underline{A} = \begin{bmatrix} -\dfrac{X^1(t) - X_{i0}}{\varrho_{i0}^1(t)} & -\dfrac{Y^1(t) - Y_{i0}}{\varrho_{i0}^1(t)} & -\dfrac{Z^1(t) - Z_{i0}}{\varrho_{i0}^1(t)} & -c \\[2ex] -\dfrac{X^2(t) - X_{i0}}{\varrho_{i0}^2(t)} & -\dfrac{Y^2(t) - Y_{i0}}{\varrho_{i0}^2(t)} & -\dfrac{Z^2(t) - Z_{i0}}{\varrho_{i0}^2(t)} & -c \\[2ex] -\dfrac{X^3(t) - X_{i0}}{\varrho_{i0}^3(t)} & -\dfrac{Y^3(t) - Y_{i0}}{\varrho_{i0}^3(t)} & -\dfrac{Z^3(t) - Z_{i0}}{\varrho_{i0}^3(t)} & -c \\[2ex] -\dfrac{X^4(t) - X_{i0}}{\varrho_{i0}^4(t)} & -\dfrac{Y^4(t) - Y_{i0}}{\varrho_{i0}^4(t)} & -\dfrac{Z^4(t) - Z_{i0}}{\varrho_{i0}^4(t)} & -c \end{bmatrix}$$

$$(9.144)$$

where the first three elements in each row define the unit vectors $\Delta\underline{\varrho}^j$, $j = 1, 2, 3, 4$ pointing from the four satellites to the observing site. The solution fails if the design matrix is singular or, equivalently, its determinant becomes zero. The determinant is proportional to the scalar triple product

$$\left((\Delta\underline{\varrho}^4 - \Delta\underline{\varrho}^1),\, (\Delta\underline{\varrho}^3 - \Delta\underline{\varrho}^1),\, (\Delta\underline{\varrho}^2 - \Delta\underline{\varrho}^1) \right)$$

which can geometrically be interpreted by the volume of a body. This body is formed by the intersection points of the site-satellite vectors with the unit sphere centered at the observing site. The larger the volume of this body, the better the satellite geometry. Since good geometry should mirror a low DOP value, the reciprocal value of the volume of the geometric body is directly proportional to DOP, cf. for example Milliken and Zoller (1980). The critical configuration is given when the body degenerates to a plane. This is the case when the vectors $\Delta\underline{\varrho}^j$ form a cone with the observing site as apex, cf. Wunderlich (1992).

More generally, DOP can be calculated from the inverse of the normal equation matrix of the solution. The cofactor matrix $\underline{Q}_X$ follows from

$$\underline{Q}_X = (\underline{A}^T \underline{A})^{-1}. \qquad (9.145)$$

In this case the weight matrix has been assumed to be a unit matrix. The cofactor matrix Q_X is a 4×4 matrix where three components are contributed by the site position X, Y, Z and one component by the receiver clock. Denoting the elements of the cofactor matrix as

$$
\underline{Q}_X = \begin{bmatrix} q_{XX} & q_{XY} & q_{XZ} & q_{Xt} \\ q_{XY} & q_{YY} & q_{YZ} & q_{Yt} \\ q_{XZ} & q_{YZ} & q_{ZZ} & q_{Zt} \\ q_{Xt} & q_{Yt} & q_{Zt} & q_{tt} \end{bmatrix}, \tag{9.146}
$$

the diagonal elements are used for the following DOP definitions:

$$
\begin{aligned}
\text{GDOP} &= \sqrt{q_{XX} + q_{YY} + q_{ZZ} + q_{tt}} &&\dots && \text{geometric dilution of precision} \\
\text{PDOP} &= \sqrt{q_{XX} + q_{YY} + q_{ZZ}} &&\dots && \text{position dilution of precision} \\
\text{TDOP} &= \sqrt{q_{tt}} &&\dots && \text{time dilution of precision.}
\end{aligned}
$$

$$\tag{9.147}$$

It should be noted that the previous DOP explanation using the geometric body refers to GDOP.

These definitions deserve brief explanations in order to avoid confusion. Quite often the elements under the square root are displayed as quadratic terms. This depends on the designation of the elements of the cofactor matrix. Here, the diagonal elements of the cofactor matrix have been denoted as q_{XX}, q_{YY}, q_{ZZ} and q_{tt}, thus no superscripts appear in the DOP definitions. If the diagonal elements are denoted as e.g. q_X^2, q_Y^2, q_Z^2, q_t^2, then of course the superscripts also appear in the DOP definitions. The general rule used here is: when computing DOP values, the elements of the cofactor matrix are not squared (e.g., for GDOP the square root of the trace must be calculated).

The DOPs in (9.147) are expressed in the equatorial ECEF system. When the topocentric local coordinate system with its axes along the local north, east and vertical is used, the global cofactor matrix $\underline{Q}_X$ must be transformed into the local cofactor matrix Q_x by the law of covariance propagation. Denoting now as $\underline{Q}_X$ that part of the cofactor matrix that contains the geometrical components (disregarding the time-correlated components), the transformation is

$$\underline{Q}_x = \underline{R}\,\underline{Q}_X\,\underline{R}^T = \begin{bmatrix} q_{xx} & q_{xy} & q_{xh} \\ q_{xy} & q_{yy} & q_{yh} \\ q_{xh} & q_{yh} & q_{hh} \end{bmatrix} \tag{9.148}$$

where the rotation matrix $\underline{R} = [\underline{i},\ \underline{j},\ \underline{k}]^T$ contains the axes of the local coordinate system as given in Eq. (7.2).

In addition to the local PDOP, two additional DOP definitions are given. HDOP, the dilution of precision in the horizontal position, and VDOP, denoting the corresponding value for the vertical component, the height:

$$\begin{aligned} \text{HDOP} &= \sqrt{q_{xx} + q_{yy}} &&\cdots && \text{horizontal dilution of precision} \\ \text{VDOP} &= \sqrt{q_{hh}} &&\cdots && \text{vertical dilution of precision.} \end{aligned} \tag{9.149}$$

The discussion, so far, involves only single epoch point positioning. When designing a survey, it is helpful to know DOP for the full (observation) session. The procedure is to compute the DOP values on an epoch by epoch basis for the desired period. The time increment between epochs can be matched to the specific purposes of the planned survey. In Fig. 9.5, PDOP is shown for an entire day and was computed for 15-minute increments. No measured data are necessary to calculate DOP values! The satellite positions

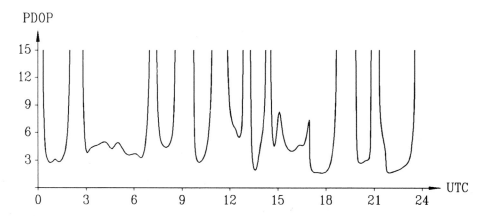

Fig. 9.5. PDOP for Graz, Austria, on April 15, 1991 (cut off elevation: 20 degrees)

can be calculated either from an almanac file or from an appropriate orbit file. Note that DOP computations are not restricted to point positioning but can also be applied to relative positioning. Starting with the design matrix for a baseline vector determination, the cofactor matrix can be computed. These DOP values can be considered as relative DOP values.

The purposes of DOP are twofold. First, it is useful in planning a survey and, secondly, it may be helpful in interpreting processed baseline vectors. For example, data with poor DOP could possibly be omitted.

Finally, the correlation of DOP with the positioning accuracy is considered. Denoting the measurement accuracy by σ_0 (i.e., the standard deviation), the positioning accuracy σ follows from the product of DOP and measurement accuracy, thus

$$\sigma = \text{DOP} \, \sigma_0 \, . \tag{9.150}$$

Applying this formula to the specific DOP definitions,

$\text{GDOP} \, \sigma_0 \quad \ldots \quad$ geometric accuracy in position and time

$\text{PDOP} \, \sigma_0 \quad \ldots \quad$ accuracy in position

$\text{TDOP} \, \sigma_0 \quad \ldots \quad$ accuracy in time

$\text{HDOP} \, \sigma_0 \quad \ldots \quad$ accuracy in the horizontal position

$\text{VDOP} \, \sigma_0 \quad \ldots \quad$ accuracy in vertical direction

results, cf. Wells et al. (1987). The list of DOP definitions is not restricted to those given here. The meaning of other DOP definitions can be derived from the associated acronyms.

10. Transformation of GPS results

10.1 Introduction

The reference frame of GPS is the World Geodetic System 1984 (WGS-84), cf. Decker (1986). When using GPS, the coordinates of terrestrial sites for example are obtained in the same reference frame. However, the surveyor is not, usually, interested in computing the coordinates of the terrestrial points in a global frame. Rather, the results are preferred in a local coordinate frame either as geodetic (i.e., ellipsoidal) coordinates, as plane coordinates, or as vectors combined with other terrestrial data. Since the WGS-84 is a geocentric system and the local system usually is not, certain transformations are required. The subsequent sections deal with the transformations most frequently used.

10.2 Coordinate transformations

10.2.1 Cartesian coordinates and ellipsoidal coordinates

Denoting the Cartesian (rectangular) coordinates of a point in space by X, Y, Z and assuming an ellipsoid of revolution with the same origin as the Cartesian coordinate system, the point can also be expressed by the ellipsoidal coordinates φ, λ, h, see Fig. 10.1. The relation between the Cartesian coordinates and the ellipsoidal coordinates, given in Eq. (3.6), is:

$$
\begin{aligned}
X &= (N + h) \cos \varphi \cos \lambda \\
Y &= (N + h) \cos \varphi \sin \lambda \\
Z &= \left(\frac{b^2}{a^2} N + h \right) \sin \varphi,
\end{aligned}
\tag{10.1}
$$

with N the radius of curvature in prime vertical

$$
N = \frac{a^2}{\sqrt{a^2 \cos^2 \varphi + b^2 \sin^2 \varphi}},
\tag{10.2}
$$

and a, b are the semiaxes of the reference ellipsoid. Recall that the Cartesian coordinates related to WGS-84 are also denoted ECEF coordinates and that

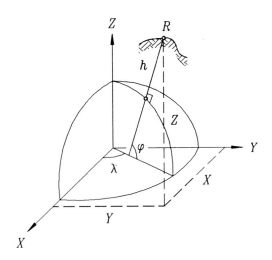

Fig. 10.1. Cartesian coordinates X, Y, Z and
ellipsoidal coordinates φ, λ, h

the origins of the ECEF coordinate system and of the WGS-84 ellipsoid of
revolution are identical (i.e., geocentric).

The formulas (10.1) transform ellipsoidal coordinates φ, λ, h into Cartesian coordinates X, Y, Z. For GPS applications, the inverse transformation
is more important since the Cartesian coordinates are given and the ellipsoidal coordinates sought. Thus, the task is now to compute the ellipsoidal
coordinates φ, λ, h from the Cartesian coordinates X, Y, Z. Usually, this
problem is solved iteratively although a solution in closed form is possible.
From X and Y the radius of a parallel,

$$p = \sqrt{X^2 + Y^2} = (N + h)\cos\varphi,\tag{10.3}$$

can be computed. This equation is rearranged as

$$h = \frac{p}{\cos\varphi} - N\tag{10.4}$$

so that the ellipsoidal height appears explicitly. Introducing by

$$e^2 = \frac{a^2 - b^2}{a^2}\tag{10.5}$$

the first numerical eccentricity, it follows $b^2/a^2 = 1 - e^2$ which can be substituted into the equation for Z in (10.1). The result

$$Z = (N + h - e^2 N)\sin\varphi\tag{10.6}$$

can be written as

$$Z = (N + h) \left(1 - e^2 \frac{N}{N + h}\right) \sin \varphi \tag{10.7}$$

equivalently. Dividing this expression by Eq. (10.3) gives

$$\frac{Z}{p} = \left(1 - e^2 \frac{N}{N + h}\right) \tan \varphi \tag{10.8}$$

which yields

$$\tan \varphi = \frac{Z}{p} \left(1 - e^2 \frac{N}{N + h}\right)^{-1}. \tag{10.9}$$

For the longitude λ the equation

$$\tan \lambda = \frac{Y}{X} \tag{10.10}$$

is obtained from Eq. (10.1) by dividing the first and the second equation.

The longitude can be directly computed from Eq. (10.10). The height h and the latitude φ are determined by Eqs. (10.4) and (10.9). The problem with (10.4) is that it depends on the (unknown) latitude. Equation (10.9) is even worse because the desired latitude is implicitly contained in the right-hand side in N. Based on these three equations, a solution can be found iteratively by the following steps:

1. Compute $p = \sqrt{X^2 + Y^2}$.

2. Compute an approximate value $\varphi_{(0)}$ from
   $$\tan \varphi_{(0)} = \frac{Z}{p}(1 - e^2)^{-1}.$$

3. Compute an approximate value $N_{(0)}$ from
   $$N_{(0)} = \frac{a^2}{\sqrt{a^2 \cos^2 \varphi_{(0)} + b^2 \sin^2 \varphi_{(0)}}}.$$

4. Compute the ellipsoidal height by
   $$h = \frac{p}{\cos \varphi_{(0)}} - N_{(0)}.$$

5. Compute an improved value for the latitude by
   $$\tan \varphi = \frac{Z}{p} \left(1 - e^2 \frac{N_{(0)}}{N_{(0)} + h}\right)^{-1}.$$

6. Check for another iteration step: if $\varphi = \varphi_{(0)}$ then go to the next step otherwise set $\varphi_{(0)} = \varphi$ and continue with step 3.

7. Compute the longitude λ from
$$\tan \lambda = \frac{Y}{X}.$$

The formulas in closed form for the transformation of X, Y, Z into φ, λ, h are

$$\varphi = \arctan \frac{Z + e'^2\, b\, \sin^3\theta}{p - e^2\, a\, \cos^3\theta}$$

$$\lambda = \arctan \frac{Y}{X} \tag{10.11}$$

$$h = \frac{p}{\cos\varphi} - N$$

where

$$\theta = \arctan \frac{Z\, a}{p\, b} \tag{10.12}$$

is an auxiliary quantity and

$$e'^2 = \frac{a^2 - b^2}{b^2} \tag{10.13}$$

is the second numerical eccentricity. Actually, there is no reason why these formulas are less popular than the iterative procedure. Either method works equally well and can be easily programmed.

10.2.2 Ellipsoidal coordinates and plane coordinates

In contrast to the previous section, only points on the ellipsoid are considered. Thus, ellipsoidal latitute φ and longitude λ are of interest here. The objective is mapping a point φ, λ on the ellipsoid into a point x, y on a plane.

There are many kinds of map projections, some being more popular than others. In principle,

$$x = x(\varphi, \lambda; a, b)$$
$$y = y(\varphi, \lambda; a, b) \tag{10.14}$$

is the general formulation of the desired map projection. Geodetic applications require conformal projections be used. Conformality means that an angle on the ellipsoid is preserved after mapping it into the plane. More precisely, the angle included by two geodesics on the ellipsoid is preserved if the two geodesics are conformally mapped into the plane. The drawback is, the geodesics on the ellipsoid mapped into the plane are in general no longer geodesics (i.e., straight lines) but curved lines.

The most important conformal mappings of the ellipsoid into the plane are:

- Conical projection. Considering the Lambert conformal projection, the cone is tangent to the ellipsoid at the standard parallel, cf. Richardus and Adler (1972). After development of the conical surface, the meridians are straight lines converging at a point called apex. This point is the center of the parallels which are projected as circular lines.

- Cylindrical projection. This is a special case of the conical projection if the apex is moved to infinity. The cylindrical surface is tangent at the equator. The two most important projections are the Transverse Mercator projection and the Universal Transverse Mercator system (UTM). More details on these two mappings are given below.

- Polar stereographic projection. This is also a special case of the Lambert projection. Let the apex of the cone move into the pole, then the cone becomes a plane. In contrast to the previously described methods (where the ellipsoid is directly transformed to the projection surface which is a plane after developing), the stereographic projection is an indirect method. The first step is a transformation of the ellipsoid to a sphere. The second step performs the transition from the sphere to the projection surface.

Transverse Mercator projection. This mapping method is also referred to as Gauss-Krüger projection. The ellipsoid is partitioned into 120 zones of 3° longitude each. In the middle of each zone there is the central meridian. The projection cylinder is transverse to the ellipsoid and tangent along a central meridian. The developing cylinder, that is opened to make a plane, yields a straight line for the central meridian with no scale distortion. In a plane, the mapped central meridian represents the y-axis (north direction). The x-axis is the mapping of the equator. In the mapping of a zone, the central meridian and the equator are special cases since all other meridians and parallels are mapped as curved lines. Because of conformality, the mapped images of the meridians and parallels are orthogonal to each other.

The numbering of the zones is related either to Greenwich or to Ferro. The latter is situated 17°40′ W with respect to Greenwich.

The mapping formulas for the Transverse Mercator projection (Gauss-Krüger projection) arise from series expansions and are given by

$$y = B(\varphi) + \frac{1}{2} N \cos^2\varphi \, t l^2$$
$$+ \frac{1}{24} N \cos^4\varphi \, t (5 - t^2 + 9\eta^2) l^4 + \ldots$$
$$\tag{10.15}$$
$$x = N \cos\varphi \, l + \frac{1}{6} N \cos^3\varphi (1 - t^2 + \eta^2) l^3$$
$$+ \frac{1}{120} N \cos^5\varphi (5 - 18 t^2 + t^4) l^5 + \ldots$$

where

$B(\varphi)$	...	arc length of meridian from equator
N	...	radius of curvature in prime vertical
$t = \tan\varphi$	...	auxiliary quantity
λ_0	...	longitude of the central meridian
$l = \lambda - \lambda_0$	...	longitude difference
$\eta^2 = e'^2 \cos^2\varphi$	...	auxiliary quantity

are used. On purpose y is given first and then x because the pair of coordinates (y, x) corresponds to (φ, λ). The arc length of meridian $B(\varphi)$ is the ellipsoidal distance from the equator to the point to be mapped and can be computed by the following series expansion:

$$B(\varphi) = \alpha \left[\varphi + \beta \sin 2\varphi + \gamma \sin 4\varphi + \delta \sin 6\varphi + \ldots\right] \tag{10.16}$$

where

$$\alpha = \frac{a + b}{2} \left(1 + \frac{1}{4} n^2 + \frac{1}{64} n^4 + \ldots\right)$$
$$\beta = -\frac{3}{2} n + \frac{9}{16} n^3 - \frac{3}{32} n^5 + \ldots$$
$$\tag{10.17}$$
$$\gamma = \frac{15}{16} n^2 - \frac{15}{32} n^4 + \ldots$$
$$\delta = -\frac{35}{48} n^3 + \frac{105}{256} n^4 - \ldots$$

and

$$n = \frac{a - b}{a + b}. \tag{10.18}$$

The inverse problem is the transformation of a point x, y in the plane to a point φ, λ on the ellipsoid. The principal methodology is the same, the coordinates on the ellipsoid are obtained from series expansions:

$$\varphi = \varphi_f - \frac{t_f}{2N_f^2}(1 + \eta_f^2)\,x^2 + \frac{t_f}{24N_f^4}(5 + 3t_f^2 + 6\eta_f^2 - 6t_f^2\,\eta_f^2)\,x^4 \ldots$$

$$l = \frac{1}{N_f \cos\varphi_f}\,x - \frac{1}{6N_f^3 \cos\varphi_f}(1 + 2t_f^2 + \eta_f^2)\,x^3 \ldots$$

$$(10.19)$$

where l is the difference in longitude with respect to the central meridian. The terms with the subscript f must be calculated based on the footpoint latitude φ_f. For the footpoint latitude, the series expansion

$$\varphi_f = \bar{y} + \beta \sin 2\bar{y} + \gamma \sin 4\bar{y} + \delta \sin 6\bar{y} \qquad (10.20)$$

is used where

$$\bar{y} = \frac{y}{\alpha}. \qquad (10.21)$$

The coefficients α, β, γ, δ depend on the ellipsoidal parameters. As an example,

$$\alpha = 6\,366\,742.5205 \text{ m}$$
$$\beta = 2.51127324 \cdot 10^{-6}$$
$$\gamma = 3.67879 \cdot 10^{-6} \qquad\qquad (10.22)$$
$$\delta = 7.38 \cdot 10^{-9}$$

are the parameters for the Bessel ellipsoid.

Universal Transverse Mercator system. The Universal Transverse Mercator system (UTM) is a modification of Transverse Mercator system. First, the ellipsoid is partitioned into 60 zones with an width of 6° longitude each. Second, a scale factor of 0.9996 is applied to the central meridian. The reason for this factor is to avoid fairly large distortions in the outer areas of a zone.

The zone numbering starts with $M1$ for the central meridian $\lambda_0 = 177°$ W and continues with $M2$ for the central meridian $\lambda_0 = 171°$ W. Thus, the central meridian $\lambda_0 = 3°$ W would be the zone $M30$.

10.2.3   Height transformation

In the previous section a point φ, λ on the ellipsoid was mapped into the plane. The ellipsoidal height could be completely ignored. In this section the primary interest is the height. The formula

$$h = H + N \tag{10.23}$$

where

$$
\begin{array}{lll}
h & \ldots & \text{ellipsoidal height} \\
H & \ldots & \text{orthometric height} \\
N & \ldots & \text{geoidal height (undulation)}
\end{array}
\tag{10.24}
$$

is the relationship between the ellipsoid and the geoid. As shown in Fig. 10.2, this formula is an approximation but is sufficiently accurate for all practical purposes. The angle ε expresses the deflection of the vertical between the slightly curved plumb line and the ellipsoidal normal. This angle does not exceed 30 arcseconds in most areas.

GPS point positioning results in X, Y, Z coordinates. After application of the transformation (10.11), the ellipsoidal heights are available. If, additionally, one of the two remaining terms in (10.23) is given, the other one can be calculated. Thus, if the geoid is known, then orthometric heights can be derived. On the other hand, if orthometric heights are known, then geoidal heights can be derived.

A short introduction is given below for readers not very experienced with ellipsoid and geoid.

Ellipsoid. An approximation of the earth's surface is an ellipsoid of revolution. This surface is formed by choosing a proper sized ellipse and rotating it about its minor axis. The earth ellipsoid is a convenient mathematical surface which has been subdivided by latitude and longitude values. Math-

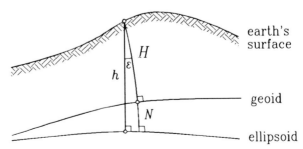

Fig. 10.2. Definition of heights

ematical distances and azimuths can be computed on this ellipsoid to millimeter accuracy and hundredths of an arcsecond. The real earth's surface only approximates the ellipsoid. As long as geodesists were performing horizontal surveys (e.g., triangulation) separate from vertical surveys (leveling), the difference between the ellipsoid and the true earth's surface (or the geoid) did not play a significant role.

This changed with the advent of space geodesy (TRANSIT, GPS) which yields both horizontal and vertical information. Now, the geometrically defined (i.e., ellipsoidal) heights obtained from space geodesy must be combined with physically defined (i.e., orthometric) heights. The ellipsoidal heights refer to the ellipsoid, and the orthometric heights to the geoid. The difference between these two surfaces will undoubtedly occupy geodesists for the next several years (maybe decades), but for the present, surveyors must learn to correct GPS heights (i.e., ellipsoidal heights) to obtain useable elevations.

There must be a distinction made between locally best-fitting (nongeocentric) ellipsoids and a global (geocentric) ellipsoid. Different local ellipsoids are primarily used because certain sized ellipsoids fit a particular portion of the earth better than a global ellipsoid. In fact, the ellipsoids chosen for an area – for example the Clarke ellipsoid for the former North American Datum 1927 (NAD-27) – were picked to minimize the difference between that ellipsoid and the geoid in that area. The center of a local ellipsoid does not coincide with the center of the true earth, but is displaced from the true center by up to some 100 meters.

Fortunately, only one global ellipsoid is used today. When it became possible to connect the datums of the various continents, geodesists decided to choose a single global reference ellipsoid. The present Geodetic Reference System 1980 (GRS-80) was chosen after much discussion. The center of the GRS-80 ellipsoid now coincides with the true center of the earth, and its surface provides an average fit of the geoid. The fit of this new surface to the geoid (often called sea level surface) is not an improvement for all places on the earth. For example, the old Clarke ellipsoid surface only differs by small amounts from the geoid surface for the U.S.; whereas, the new GRS-80 ellipsoid is approximately 30 m lower than the geoid in the eastern U.S. Surveyors using the old NAD-27 datum were accustomed to reduce their distance measurements to "sea level" when actually they were correcting lengths to map them onto the ellipsoid. Now, they must add approximately 30 m to the elevation to obtain the correct height to use for the reduction to the ellipsoid.

Geoid. The geoid is defined physically and is the surface that is used to represent the actual shape of the earth. Drawings of the geoid show it as a bumpy surface with hills and valleys similar to a topographic model.

The center of the geoid coincides with the true center of the earth and its surface is an equipotential surface. The geoid can be visualized by imagining that the earth were completely covered by water. This water surface would (in theory) be an equipotential surface since the water would flow to compensate for any height difference that would occur. In actuality, the sea level differs slightly from a true equipotential surface due to "bumps" formed by different temperatures, ocean currents, etc.

The geoid is the surface chosen for leveling datums. In many countries, the geoid that most nearly coincides with the average sea level value (measured by tide gauges) is chosen as the zero elevation. Because of this choice, the datum is often referred to as the sea level datum. This does not mean, however, that the zero elevation will coincide with the average level of sea level along a coastline. In fact, differences of about a meter exist between the zero elevation and the average sea level at any given point. The average sea level at a given point along the coast is affected by many factors so that average sea levels along the coast do not form an equipotential surface.

The geoidal surface is very irregular and it is virtually impossible to realize an exact mathematical model for the geoid. However, geodesists have developed a good approximation of the geoid through the use of spherical harmonics, cf. (4.40). The most popular mathematical model of the continental U.S. geoid was produced by R. Rapp of Ohio State University. Similarly, H. Sünkel of Graz University of Technology and others have produced geoid models for European countries.

10.3   Similarity transformations

The coordinate transformations in the previous section dealt with the transition from one type of coordinates to another type of coordinates for the same point. Cartesian coordinates X, Y, Z have been transformed into ellipsoidal coordinates φ, λ, h and two-dimensional ellipsoidal surface coordinates φ, λ have been transformed into plane coordinates x, y. Finally, the ellipsoidal height is transformed either to the orthometric height or to the geoidal height.

A similarity transformation transforms one coordinate system from a certain type to another coordinate system of the same type. For a grouping, three-dimensional, two-dimensional, and one-dimensional similarity transformations are distinguished.

10.3.1   Three-dimensional transformation

Consider two sets of three-dimensional Cartesian coordinates forming the vectors $\underline{X}$ and $\underline{X}_T$, cf. Fig. 10.3. The transformation between the two sets can be formulated by the relation

$$\underline{X}_T = \underline{c} + \mu \, \underline{R} \, \underline{X} \tag{10.25}$$

which is denoted as Helmert transformation. The term μ is a scale factor, $\underline{c}$ is the shift vector

$$\underline{c} = \begin{bmatrix} c_1 \\ c_2 \\ c_3 \end{bmatrix}, \tag{10.26}$$

and $\underline{R}$ is a rotation matrix which is composed of three successive rotations

$$\underline{R} = \underline{R}_3\{\alpha_3\} \, \underline{R}_2\{\alpha_2\} \, \underline{R}_1\{\alpha_1\} \tag{10.27}$$

and is given by

$$\underline{R} = \begin{bmatrix} \cos\alpha_2 \cos\alpha_3 & \begin{array}{c}\cos\alpha_1 \sin\alpha_3 \\ + \sin\alpha_1 \sin\alpha_2 \cos\alpha_3\end{array} & \begin{array}{c}\sin\alpha_1 \sin\alpha_3 \\ - \cos\alpha_1 \sin\alpha_2 \cos\alpha_3\end{array} \\[2ex] -\cos\alpha_2 \sin\alpha_3 & \begin{array}{c}\cos\alpha_1 \cos\alpha_3 \\ - \sin\alpha_1 \sin\alpha_2 \sin\alpha_3\end{array} & \begin{array}{c}\sin\alpha_1 \cos\alpha_3 \\ + \cos\alpha_1 \sin\alpha_2 \sin\alpha_3\end{array} \\[2ex] \sin\alpha_2 & -\sin\alpha_1 \cos\alpha_2 & \cos\alpha_1 \cos\alpha_2 \end{bmatrix}$$

$$\tag{10.28}$$

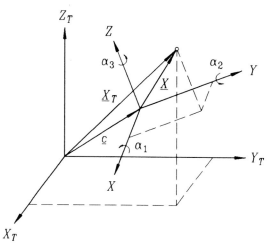

Fig. 10.3. Three-dimensional similarity transformation

after substitution of the single rotation matrices, cf. Eq. (3.8) accordingly.

Before proceeding, the seven parameter Helmert similarity transformation, cf. Eq. (10.25), will be briefly discussed. The components of the translation vector $\underline{c}$ account for the coordinates of the origin of the $\underline{X}$ system in the $\underline{X}_T$ system. Usually, a single scale factor μ is considered. More generally (but with GPS not necessary), three scale factors, one for each axis, could be used. The rotation matrix $\underline{R}$ is an orthogonal matrix with the three parameters α_i as unknowns.

In the case of known transformation parameters $\underline{c}$, μ, $\underline{R}$, a point from the $\underline{X}$ system could be transformed into the $\underline{X}_T$ system by (10.25).

If the transformation parameters are unknown, they can be determined with the aid of common (identical) points. This means that the coordinates of the same point are given in both systems. Since each common point (given by $\underline{X}_T$ and $\underline{X}$) yields three equations, two common points and one additional common component (e.g., height) are sufficient to solve for the seven unknown parameters. In practice, redundant common point information is used and the unknown parameters are calculated by an adjustment procedure.

Since in Eq. (10.25) the parameters are mixed nonlinearly, a linearization must be performed. Symbolizing known approximations by parentheses

$$(\underline{c}) \qquad (\mu) \qquad (\underline{R}), \tag{10.29}$$

the final (adjusted) values are obtained by

$$\underline{c} = (\underline{c}) + d\underline{c}$$

$$\mu = (\mu)[1 + d\mu] \tag{10.30}$$

$$\underline{R} = d\underline{R}\,(\underline{R})\,.$$

Along with $d\mu$, the increments for the translation vector

$$d\underline{c} = \begin{bmatrix} dc_1 \\ dc_2 \\ dc_3 \end{bmatrix} \tag{10.31}$$

and the elements (i.e., increments for the rotation angles) of the differential rotation matrix $d\underline{R}$

$$d\underline{R} = \begin{bmatrix} 1 & d\alpha_3 & -d\alpha_2 \\ -d\alpha_3 & 1 & d\alpha_1 \\ d\alpha_2 & -d\alpha_1 & 1 \end{bmatrix} \tag{10.32}$$

are now the new unknowns. The differential rotation matrix is obtained by introducing the small quantities $d\alpha_i$ into Eq. (10.28), setting $\cos d\alpha_i \approx 1$ and $\sin d\alpha_i \approx d\alpha_i$, and neglecting second-order terms.

Skipping the details which can be found for example in Hofmann-Wellen-hof (1990), the linearized model for a single point i is formulated as

$$\underline{X}_{T_i} = (\underline{X}_T)_i + \underline{A}_i\, d\underline{p} \tag{10.33}$$

where

$$(\underline{X}_T)_i = (\mu)(\underline{R})\,\underline{X}_i + (\underline{c}) \tag{10.34}$$

which can be calculated from the approximate transformation parameters and the given coordinates $\underline{X}_i$. The design matrix $\underline{A}_i$ and the parameter vector $d\underline{p}$ are:

$$\underline{A}_i = \begin{bmatrix} 1 & 0 & 0 & (\Delta X_i) & 0 & -(\Delta Z_i) & (\Delta Y_i) \\ 0 & 1 & 0 & (\Delta Y_i) & (\Delta Z_i) & 0 & -(\Delta X_i) \\ 0 & 0 & 1 & (\Delta Z_i) & -(\Delta Y_i) & (\Delta X_i) & 0 \end{bmatrix} \tag{10.35}$$

$$d\underline{p} = \begin{bmatrix} dc_1 & dc_2 & dc_3 & d\mu & d\alpha_1 & d\alpha_2 & d\alpha_3 \end{bmatrix}^T.$$

The components (ΔX_i), (ΔY_i), (ΔZ_i) in the design matrix are abbreviations for

$$(\Delta X_i) = (X_T)_i - (c_1)$$
$$(\Delta Y_i) = (Y_T)_i - (c_2) \tag{10.36}$$
$$(\Delta Z_i) = (Z_T)_i - (c_3)$$

where the components of the approximate vector $(\underline{X}_T)_i$ are obtained from (10.34).

Equation (10.33) in combination with (10.34) and (10.35) is now a system of linear equations for point i. For n common points, the design matrix A is:

$$\underline{A} = \begin{bmatrix} \underline{A}_1 \\ \underline{A}_2 \\ \vdots \\ \underline{A}_n \end{bmatrix}. \tag{10.37}$$

In detail, for three common points the design matrix is

$$\underline{A} = \begin{bmatrix} 1 & 0 & 0 & (\Delta X_1) & 0 & -(\Delta Z_1) & (\Delta Y_1) \\ 0 & 1 & 0 & (\Delta Y_1) & (\Delta Z_1) & 0 & -(\Delta X_1) \\ 0 & 0 & 1 & (\Delta Z_1) & -(\Delta Y_1) & (\Delta X_1) & 0 \\ 1 & 0 & 0 & (\Delta X_2) & 0 & -(\Delta Z_2) & (\Delta Y_2) \\ 0 & 1 & 0 & (\Delta Y_2) & (\Delta Z_2) & 0 & -(\Delta X_2) \\ 0 & 0 & 1 & (\Delta Z_2) & -(\Delta Y_2) & (\Delta X_2) & 0 \\ 1 & 0 & 0 & (\Delta X_3) & 0 & -(\Delta Z_3) & (\Delta Y_3) \\ 0 & 1 & 0 & (\Delta Y_3) & (\Delta Z_3) & 0 & -(\Delta X_3) \\ 0 & 0 & 1 & (\Delta Z_3) & -(\Delta Y_3) & (\Delta X_3) & 0 \end{bmatrix} \tag{10.38}$$

which leads to a slightly overdetermined system. The adjustment via the
normal equations yields the parameter vector dp and the adjusted values by
(10.30). Once the seven parameters of the Helmert similarity transformation
are determined, formula (10.25) can be used to transform other points. For
example, the WGS-84 coordinates of a point obtained from GPS observations
could be transformed to a nongeocentric local national system.

10.3.2 Two-dimensional transformation

Here, two-dimensional coordinates are considered. The two different sets of
plane coordinates are implicit in the vectors $\underline{x} = (x, y)^T$ and $\underline{x}_T = (x_T, y_T)^T$,
cf. Fig. 10.4. The two-dimensional similarity transformation is defined by

$$\underline{x}_T = \underline{c} + \mu\,\underline{R}\,\underline{x} \tag{10.39}$$

with a scale factor μ, the shift vector

$$\underline{c} = \begin{bmatrix} c_1 \\ c_2 \end{bmatrix} \tag{10.40}$$

and the rotation matrix

$$\underline{R} = \begin{bmatrix} \cos\alpha & -\sin\alpha \\ \sin\alpha & \cos\alpha \end{bmatrix} \tag{10.41}$$

which comprises a single rotation angle. Hence, Eq. (10.39) in combination
with (10.40) and (10.41) is the two-dimensional Helmert transformation with

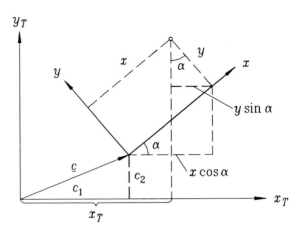

Fig. 10.4. Two-dimensional similarity transformation

four parameters: the two translation components c_1, c_2, the scale factor μ, and the rotation angle α. Substituting (10.40) and (10.41) into (10.39) yields the single components of the transformed point:

$$x_T = c_1 + \mu\, x\, \cos\alpha - \mu\, y\, \sin\alpha$$
$$y_T = c_2 + \mu\, x\, \sin\alpha + \mu\, y\, \cos\alpha . \tag{10.42}$$

These simple formulas can be verified by the geometry given in Fig. 10.4 where the contributing terms for x_T are indicated.

When the transformation parameters $\underline{c}$, μ, $\underline{R}$ are known, a point in the $\underline{x}$ system can be transformed into the $\underline{x}_T$ system by (10.39).

When the transformation parameters are unknown, they can be determined – analogously to the three-dimensional case – using common points. Two common points, each yields two equations, are sufficient to solve for the four unknown parameters. In practice, redundant common point information is used and the unknown parameters are calculated by least squares adjustment.

As seen from (10.42), the unknowns appear again in nonlinear form. However, using the auxiliary unknowns

$$p = \mu\, \cos\alpha$$
$$q = \mu\, \sin\alpha , \tag{10.43}$$

linear equations

$$x_T = c_1 + p\,x - q\,y$$
$$y_T = c_2 + q\,x + p\,y \tag{10.44}$$

with respect to the unknowns are obtained. In the case of redundant common points, the solution is obtained by the conventional least squares adjustment procedure. The scale factor and the rotation angle are then determined from the auxiliary unknowns by

$$\mu = \sqrt{p^2 + q^2}$$
$$\tan\alpha = \frac{q}{p} . \tag{10.45}$$

10.3.3 One-dimensional transformation

One of the distinct and modern features of GPS is that three-dimensional (3D) coordinates are obtained in a common frame. There is no separation between the horizontal coordinates and the height of a point because all three components are calculated together by the same procedure. In classical geodesy, horizontal coordinates and heights were obtained independently.

Now, the question arises why in the previous section a two-dimensional (2D) transformation and now a one-dimensional (1D) transformation are discussed. The answer lies in the data of the past. Many countries have excellent horizontal control points available but often fairly poor ellipsoidal heights because of lack of geoidal heights. Thus, it does make sense to use the 2D transformation if height information is not available. Similarly, the 1D transformation can be used to transform heights.

Symbolically, the 1D transformation is obtained by 3D $\ominus$ 2D. In more detail, the parameters for the 1D transformation are obtained by "subtracting" the parameters of the 2D transformation from the parameters of the 3D transformation. This looks like

$$
\left.\begin{array}{lccccccc}
\text{3D} & c_1 & c_2 & c_3 & \mu & \alpha_1 & \alpha_2 & \alpha_3 \\
\text{2D} & c_1 & c_2 & & \mu & & & \alpha_3 \\
\hline
\text{1D} & & & c_3 & & \alpha_1 & \alpha_2 &
\end{array}\right\} \ominus
\tag{10.46}
$$

where for the 2D transformation the rotation angle has been equipped with the corresponding subscript. In other words, the 3D transformation for the horizontal coordinates and the height is composed of a 2D transformation for the horizontal coordinates and a 1D transformation for the heights.

Concentrating now on the 1D transformation, from the parameters in (10.46) it can be seen that the transformation consists of a shift along the vertical axis, a tilt (rotation) about the north-south axis, and a tilt about the east-west axis. These three unknowns are determined by using the height information of three common points.

Transformation using heigths. Assume that for three points in a GPS network the orthometric heights H_i (i.e., elevations) and the ellipsoidal heights h_i are known. The latter can be transformed into approximate elevations (H_i) using a geoid model. Normally, there are discrepancies between the heights H_i and (H_i) due to the combined effects of the GPS systematic errors and the errors in the geoid modeling, see Collins (1989). The mathematical model for the discrepancies is given by

$$
H_i - (H_i) = dH - y_i \, d\alpha_1 + x_i \, d\alpha_2
\tag{10.47}
$$

where dH is the vertical shift. The rotation about the x-axis and y-axis are denoted $d\alpha_1$ and $d\alpha_2$. Equation (10.47) is represented in the local coordinate frame with the point's plane coordinates x_i, y_i, and formally corresponds to the third component in the three-dimensional transformation, cf. for example Eq. (10.35). Geometrically, the model equation for the discrepancies

$H_i - (H_i)$ can be interpreted as the equation of a plane which could be extended to a higher-order surface to take into account more irregular geoid structures.

The FGCC specifications for GPS surveys require that surveys to be included into the national network be tied to a minimum of four benchmarks well-distributed geometrically throughout the project area (e.g., corners of project area). The additional benchmarks (ties) enable a least squares adjustment of the model equations (10.47) and provide the necessary check on the computation of the rotation of the ellipsoid to the geoid. A good practice is to perform the rotation using three of the elevations and to check the rotated elevation of the fourth point against the true elevation. The two values should agree within a few centimeters under normal conditions. An additional check on the correctness of the transformation is provided by inspecting the magnitude of the two rotation angles computed by the least squares adjustment program. Normally, these angles should be less than 5 arcseconds, although they can on occasion be greater.

Normally, the elevations of points in a network of small size, let us say $10\,\text{km} \times 10\,\text{km}$, can be determined with an accuracy of about $3\,\text{cm}$. In areas where the geoid is well-known or where the generalized model adequately describes the surface, much larger areas can be surveyed using this method with comparable accuracies being achieved.

Transformation using height differences. In the preceding paragraph, the importance of geoidal heights has been stressed. The ellipsoidal height determined by GPS can be transformed to the orthometric height if the geoidal height is known.

There are times, however, when it is only desired to measure changes in elevation. For example, when it is desired to measure the subsidence rate of a point (e.g., oil platform). In such cases the importance of a well-known geoid diminishes because relative heights are considered. For two points then

$$H_1 = h_1 - N_1$$
$$H_2 = h_2 - N_2 \tag{10.48}$$

are the height relations and

$$H_2 - H_1 = h_2 - h_1 - (N_2 - N_1) \tag{10.49}$$

is the height difference or height change between the points 1 and 2. Here, only the difference of the geoidal heights affects the result. Thus, if the geoidal heights are constant in a local area, meaning that the separation between the geoid and the ellipsoid is constant, they can be ignored. Similarly, if the geoid has a constant slope with the ellipsoid, the heights can be computed accurately by rotating the GPS heights into the geoidal surface, see Collins (1989).

10.4 Combining GPS and terrestrial data

This section is restricted to basic considerations; otherwise, the text would become a treatment of adjustment theory and not a GPS book. The combination of data sets from two or more different sources is strictly an adjustment problem and not all cases are covered here. Thus, only the most frequently applied combinations of GPS data and terrestrial data are investigated.

10.4.1 Data transformation

One primary task in combining GPS data with terrestrial data is the transformation of geocentric WGS-84 coordinates to local terrestrial coordinates. As mentioned earlier, the terrestrial system uses a locally best-fitting ellipsoid, e.g. the Clarke ellipsoid or the GRS-80 ellipsoid in the U.S. and the Bessel ellipsoid in many parts of Europe. The local ellipsoid is linked to a nongeocentric Cartesian coordinate system where the origin coincides with the center of the ellipsoid. Plane coordinates such as Gauss-Krüger coordinates are obtained by mapping the ellipsoid into the plane.

The GPS coordinates are denoted by the subscript "GPS", and the terrestrial coordinates referring to a local system are denoted by the subscript "LS". Thus, GPS observations provide $(X, Y, Z)_{GPS}$ coordinates. Local plane coordinates $(x, y)_{LS}$ can be transformed via formulas (10.19) to ellipsoidal coordinates $(\varphi, \lambda)_{LS}$. If orthometric and geoidal heights are available, then also ellipsoidal heights can be computed so that the coordinate triple $(\varphi, \lambda, h)_{LS}$ is obtained. These coordinates can be transformed to Cartesian coordinates $(X, Y, Z)_{LS}$ by (10.1).

The inverse transformation from $(X, Y, Z)_{LS}$ to $(x, y)_{LS}$ can be achieved by first applying (10.11) or by using an iteration procedure. The mapping of the ellipsoidal surface coordinates into the plane is performed via Eq. (10.15) where the height is not needed.

Transformation in three-dimensional space. The task being considered is to combine $(X, Y, Z)_{GPS}$ coordinates and $(x, y)_{LS}$ in three-dimensional space. Details are shown in Table 10.1 and in the following algorithm:

Table 10.1.   Transforming GPS and terrestrial data in three-dimensional space and mapping to the local plane system

WGS-84 geocentric		Local system nongeocentric	Comments
		Ellipsoid	Reference surface, e.g., Bessel, Clarke
$(X, Y, Z)_{GPS}$	common $\longleftrightarrow$ points	$(X, Y, Z)_{LS}$	Seven parameters of the Helmert transformation are determined by common points
$(X, Y, Z)_{GPS}$	7 param- $\longrightarrow$ eters	$(X, Y, Z)_{LS}$	The WGS-84 coordinates are transformed to the local system with known parameters
		$\downarrow$ $(\varphi, \lambda, h)_{LS}$	Cartesian coordinates are transformed to ellipsoidal coordinates referring to the local system
		$\downarrow$ $(\varphi, \lambda)_{LS}$	The heights are omitted, thus, surface coordinates are considered
		$\downarrow$ $(x, y)_{LS}$	Mapping of ellipsoidal surface coordinates into the plane referring to the local system

1. Transform $(x, y)_{LS}$ by the formulas (10.19) to $(\varphi, \lambda)_{LS}$.

2. To get a complete coordinate triple $(\varphi, \lambda, h)_{LS}$, ellipsoidal heights must be available.

3. Transform $(\varphi, \lambda, h)_{LS}$ into $(X, Y, Z)_{LS}$ by (10.1).

4. Common points of coordinates $(X, Y, Z)_{GPS}$ referring to the WGS-84 and of $(X, Y, Z)_{LS}$ referring to the local system are used to determine the seven parameters of the Helmert transformation (10.25).

5. Now, $(X, Y, Z)_{GPS}$ coordinates (referring to the WGS-84) for points other than common points can be transformed via Eq. (10.25) to $(X, Y, Z)_{LS}$ referring to the local system using the transformation parameters computed in the previous step.

6. All $(X, Y, Z)_{LS}$ coordinates can be transformed to $(\varphi, \lambda, h)_{LS}$ referring to the local system by (10.11) or by the iteration procedure.

7. Omitting the heights, the ellipsoidal surface coordinates $(\varphi, \lambda)_{LS}$ are mapped into the plane by (10.15) and yield $(x, y)_{LS}$ in the local plane system.

Table 10.1 is a type of flow diagram for the transformation of GPS data and terrestrial data in three-dimensional space which repeats the most important steps of the procedure described above. Note, however, that it is based on the assumption that the points in the local system being used as common points are already given as $(X, Y, Z)_{LS}$ coordinates. In the local system, a nongeocentric Cartesian coordinate system with an ellipsoid (e.g., Bessel, Clarke) as reference surface is used. For the WGS-84, the ellipsoid is intentionally omitted because for this type of transformation it is never needed.

The advantage of the three-dimensional approach is that no a priori information is required for the seven parameter Helmert transformation. This means that no information on the three shift, one scale, and three rotation parameters of the two systems is necessary. The main disadvantage of the method is that for points in the local system ellipsoidal heights (and thus geoidal heights) are required. However, as reported by Schmitt et al. (1991), the effect of incorrect heights of the common points is often negligible on the plane coordinates (x, y) and on the ellipsoidal coordinates (φ, λ). For example, incorrect heights cause a tilt of a 20 km × 20 km network by an amount of 5 m in space; however, the effect on the plane coordinates is only approximately 1 mm.

Many variations for the combination in three-dimensional space exist. Mentioning a single additional example, Schödlbauer et al. (1989) obtain equivalent results to the method described by mapping the GPS coordinates into the plane, adopting the height, and performing a three-dimensional affine transformation via common points.

Transformation in two-dimensional space. Now, the task being considered is to combine $(X, Y, Z)_{GPS}$ coordinates and $(x, y)_{LS}$ in a plane without height information. The following steps, theoretically, would provide a solution:

1. All Cartesian coordinates $(X, Y, Z)_{GPS}$ referring to the WGS-84 can be transformed to $(\varphi, \lambda, h)_{GPS}$ by Eq. (10.11). The same ellipsoidal parameters as for the local system (e.g., Bessel ellipsoid) are used.

2. Omitting the heights, $(\varphi, \lambda)_{GPS}$ is mapped into the plane by (10.15) and yields $(x, y)_{GPS}$. Note that the dimension of the ellipsoid enters again in the corresponding mapping formulas.

3. Common points, in the minimum two, are selected from the $(x, y)_{GPS}$ coordinates and from the $(x, y)_{LS}$ coordinates referring to the local system to calculate the transformation parameters of the two-dimensional Helmert transformation, cf. Eq. (10.39).

4. With known transformation parameters, Eq. (10.39) transforms the remaining points $(x, y)_{GPS}$ to the local system.

Without using height information, the data transformation works well for small sized networks, low elevations, and for areas where geoidal heights do not change. However, the coordinates $(x, y)_{GPS}$ depend on the dimension of the ellipsoid and, further, the coordinates $(x, y)_{LS}$ are influenced by the displacement of the nongeocentric local ellipsoid. These dependencies lead to a distortion of the point clusters which, in principle, is not matched by a similarity transformation, cf. Lichtenegger (1991). A numerical example is given in Schmitt et al. (1991) and shows that the use of the WGS-84 ellipsoid and the two-dimensional Helmert transformation leads to unacceptable discrepancies in the range of 8 mm to 15 mm for a network covering an area of 200 km × 200 km.

One way to overcome the problem is to apply other transformations such as the affine transformation between the two coordinate sets in the plane. Another way is to transform the $(x, y)_{LS}$ coordinates approximately to $(X, Y, Z)_{LS}$ by using approximate ellipsoidal heights. A three-dimensional similarity transformation by means of common points delivers approximate parameters which enable a transformation of all GPS coordinates into plane. The final similarity transformation is performed in two-dimensional space.

Transformation of baseline vectors. So far, the transformation of GPS coordinates into the local system has been considered. Surveyors are also interested in distances and (horizontal and vertical) angles computed from baseline vectors. As an example, denote the baseline vector between the points A and B by $\underline{b}_{AB}$, the spatial distance s_{AB}, the ellipsoidal azimuth a_{AB}, and the ellipsoidal zenith distance by z_{AB}. These quantities are obtained explicitly by the equations

$$s_{AB} = \|\underline{b}_{AB}\|$$

$$a_{AB} = \arctan \frac{\underline{b}_{AB} \cdot \underline{j}}{\underline{b}_{AB} \cdot \underline{i}}$$ (10.50)

$$z_{AB} = \arctan \frac{\sqrt{(\underline{b}_{AB} \cdot \underline{i})^2 + (\underline{b}_{AB} \cdot \underline{j})^2}}{\underline{b}_{AB} \cdot \underline{k}}$$

where the vectors $\underline{i}$, $\underline{j}$, $\underline{k}$ are the axes of the local coordinate frame according to Eq. (7.2). The quantities s_{AB}, a_{AB}, z_{AB} can be regarded as refraction-free observables where of course correlations must be taken into account.

10.4.2   Adjustment

This section is kept as brief as possible because other texts cover the numerous aspects of adjustment theory. No details are given on cofactor matrices, variances, covariances and error propagation law. A single example is chosen to show the variety of possible applications.

The formula for the three-dimensional Helmert transformation is given by Eq. (10.25)

$$\underline{X}_T = \underline{c} + \mu \, \underline{R} \, \underline{X}$$ (10.51)

and shows the interrelation between two systems. Continuing with the coordinates of a local system LS and global GPS coordinates, the relation above becomes

$$\underline{X}_{GPS} = \underline{c} + \mu \, \underline{R} \, \underline{X}_{LS} \,.$$ (10.52)

Assuming the coordinates of both systems as stochastic, a noise vector can be added to $\underline{X}_{GPS}$ and to $\underline{X}_{LS}$ as well. The relation in this form is

$$\underline{X}_{GPS} + \underline{n}_{GPS} = \underline{c} + \mu \, \underline{R} \, (\underline{X}_{LS} + \underline{n}_{LS})$$ (10.53)

and is called Gauss-Helmert model. From the adjustment point of view, it is advantageous to denote the adjusted coordinates of the local system by $\underline{\hat{X}}_{LS} = \underline{X}_{LS} + \underline{n}_{LS}$. The model is now reformulated as

$$\underline{X}_{LS} + \underline{n}_{LS} \quad = \underline{\hat{X}}_{LS}$$
$$\underline{X}_{GPS} + \underline{n}_{GPS} = \underline{c} + \mu \, \underline{R} \, \underline{\hat{X}}_{LS}$$ (10.54)

and is called Gauss-Markov model. Here, the adjusted coordinates $\underline{\hat{X}}_{LS}$ are used on the one hand as "observations" of the unknowns and on the other hand as unknowns.

The formula above is the basic model which can easily be enlarged by other (terrestrial) observations. Jäger and van Mierlo (1991) add, for example, terrestrial measurements of distances and angles. Gathering these observations in a vector $\underline{l}$ and adding the corresponding noise, the model above can be supplemented with an equation of the type $\underline{l}_{LS} + \underline{n} = \underline{l}(\hat{\underline{X}}_{LS})$.

In principle, any kind of geodetic measurement can be implemented if the integrated geodesy adjustment model is used. The basic idea is that any geodetic measurement can be expressed as a function of one or more position vectors $\underline{X}$ and of the gravity field W of the earth. The usually nonlinear function must be linearized where the gravity field W is split into the normal potential U of an ellipsoid and the disturbing potential T, thus $W = U + T$. Applying a minimum principle leads to the collocation formulas, cf. Moritz (1980), Chap. 11.

Many examples integrating GPS and other data can be found in technical publications. For example, Hein et al. (1989) integrate GPS and gravity data. Hein (1990a) computes orthometric heights by combining GPS and gravity data. Grant (1988) attempts to detect earth deformations from a GPS and terrestrial data combination. Delikaraoglou and Lahaye (1990) cite further contributions.

10.5 Fiducial point concept

Although the use of the GPS fiducial point concept is primarily to improve satellite orbits, cf. for example Ashkenazi et al. (1990), it nevertheless should be considered in the section on transformations. Fiducial points are sites whose positions are accurately known from a GPS independent method such as VLBI or SLR. The concept of fiducial points is quite simple: during a GPS campaign at least three fiducial points in the area of the campaign are also equipped with receivers (apart from the points to be determined). The geometry of the fiducial points with respect to the remaining points has a strong effect on the GPS accuracy in this region, cf. Lichten et al. (1989). The measurements of the fiducial points serve two purposes: the data are used for baseline determinations and also for orbit improvements, cf. Davidson et al. (1985).

Regarding transformations, the $(X, Y, Z)_{GPS}$ coordinates (referring to the WGS-84) can be transformed into the fiducial point system by a three-dimensional similarity transformation. In accordance with the previous paragraph, the coordinates of three fiducial points are necessary in both systems to determine the transformation parameters. Apart from the accuracy of fiducial points and the primary goal of orbit improvement, the fiducial points can also serve as common points for the transformation.

11. Software modules

11.1 Introduction

The policy used throughout this book of not endorsing commercial products is also adhered in this chapter. There are no references given, although publications of authors describing various software packages have been used. The main goal of this chapter is to provide potential GPS users with useful information concerning the various features of different software modules. The following sections cannot give a complete list of all imaginable features, but they will be a help in determining which features are worthwhile. However, the main consideration is that the software (and the corresponding receiver) should match the personal requirements. Most applications do not require all given features.

Each of the following paragraphs describing various terms is kept as brief as possible. More extensive descriptions are found in other chapters of this book. A summary table is provided for each section at the end of this chapter in Sect. 11.10 which can be used as a kind of checklist when comparing different softwares.

Most software modules are designed to run on PC type computers, although, this will undoubtedly change as better computers become available. Some software is used in the receiver, while other software operates interactively between the PC and the receiver (or with another equipment).

Table 11.1 gives an overview of the modules described in the following sections.

Table 11.1. Software modules

• Planning
• Data transfer
• Data processing
• Quality control
• Network adjustment
• Data base management
• Utilities
• Flexibility

11.2   Planning

11.2.1   Satellite visibility

In the planning stage of a survey it is often necessary to have a visibility table listing the locations of the satellites. Also, rise and set times of the satellites are useful. When the final constellation is deployed, and a 24-hour window is available, satellite visibility diagrams and tables will not be as important as today.

In addition to a visibility table a (polar) sky plot of the individual satellites is useful. A helpful option is the ability of entering site obstructions at both ends of a baseline to determine the "combined" visibility.

11.2.2   Satellite geometry

The DOP values are highly correlated with satellite geometry and, thus, the number of visible satellites. A table or a plot of DOP helps to select periods with good geometry. Since DOP is directly correlated with the achievable positioning accuracy, observation windows (especially for kinematic surveys) should be chosen with periods of good DOP.

11.2.3   Simulations

Simulations can be used to plan a survey. The distribution of the sites, the observation period, and the length of observation all contribute to the design matrix so that a cofactor matrix can be computed.

11.2.4   Receiver preprogramming

Some receivers require the preprogramming of start and stop times, and of other parameters such as sampling rate, satellite preselection, etc. Virtually all receivers, however, permit manual operation by field personnel. Newer receivers are fully automatic and do not require any operator interaction.

11.3   Data transfer

11.3.1   Downloading data

Upon completion of a survey, the data stored in each receiver must be downloaded to a computer. There are several possibilities for downloading these data depending on the storage device. One technique is to copy the data onto a cassette or floppy disk, using a cable or a radio link. Another method is to copy files from the receiver directly to a computer disk.

11.3.2 Decoding data

The downloaded data are often decoded from a special compact receiver format to a general binary format. Software that yields the navigation message (e.g., broadcast ephemerides), the measured quantities (i.e., code ranges, carrier phases, and Doppler data), and other information (e.g, field log) is used to provide files required by the processing software.

11.3.3 Checking transferred data

The transferred data should be verified for consistency and integrity. This could also be performed immediately after downloading and before decoding, but these checks would be more limited.

11.3.4 File handling

A good practice is to set up tables containing site names and sessions. Additional information such as known coordinates, observer names, and weather conditions are also useful. Control stations which will be used as fiducial points (i.e., known coordinates) should be properly designated. Care should be exercised in not mixing data from different sites or sessions. Normally, files are named to correspond to unique site and session identifiers.

11.3.5 Baseline definition

For projects where more than a single baseline vector has been measured there are different combinations of baselines that can be computed. The processing software should offer the choice of calculating single baseline vectors (with all possible combinations) or of calculating multipoint vectors (where for n sites, there are $n - 1$ independent baselines). User interaction for a reformulation of baselines must be possible.

11.4 Data processing

11.4.1 Generating RINEX formatted data

An important software feature is the ability to convert measurements into the RINEX format. Most data exchange uses this format so that data from two different types of receivers can be used to process a baseline vector.

11.4.2   Ephemerides

The broadcast ephemerides must be converted to an orbit file where the coordinates of the satellites are available for selected epochs. Satellite coordinates at other epochs are normally interpolated from a few selected epochs. In addition to the broadcast ephemerides, an option for using the precise orbit is essential for high-accuracy surveys.

11.4.3   Code data processing

Before a baseline vector can be computed, the coordinates of each endpoint of the line are normally first determined using the code pseudoranges. In addition to the 3D positions, the receiver clock offset is obtained. For stations with known coordinates, the receiver clock information is more precise.

11.4.4   Phase data processing

There are numerous options available for processing phase data. The raw phases can be used or they can be combined to form single-, double-, or triple-differences, for either or both frequencies. There is also a variety of frequency dependent combinations such as single frequency, dual frequency, ionospheric-free combination, and others. In addition, phase/code range combinations should be possible.

11.4.5   Data analysis

Reliable software should provide sophisticated data analysis. Gross errors should automatically be detected and eliminated. Automatic cycle slip detection is essential. Plot options are an excellent tool to visualize errors such as cycle slips.

11.4.6   Covariance matrices

Correct modeling of the covariances can be fairly complex for various data combinations. Software should cover the most common cases such as covariances for double-differences in both the single baseline and the multipoint mode.

11.4.7   Modeling the atmosphere

There should be various choices for modeling the atmosphere. In addition to a standard atmosphere model, various models for the ionosphere and the troposphere allow optimizing data processing, especially for long baselines.

11.4.8 Parameter estimation

Optimal software should provide flexibility in selecting the parameters to be adjusted and the way that the adjustment will be performed. The following options are some that optimal software should contain:

- Computation of coordinates (baseline vectors). This includes both static and kinematic surveys. A choice of using the single baseline or multipoint mode should be available. The ability to combine various sessions is also useful.

- Clock parameters. Usually a clock offset and a clock drift are estimated. Drift results can be expressed as polynomial coefficients.

- Ambiguities. In the first adjustment, the ambiguities are calculated as real values. Subsequently the ambiguities are fixed as integer values using different techniques. In succeeding adjustments the ambiguities are substituted as known values. The ability to bias integers one at a time can prove useful.

- Orbit. Certainly some insight is necessary to judge the efficiency of orbit computations. The result should be the six orbital elements of the satellites and the corresponding partial derivatives (osculating ellipses).

11.5 Quality control

11.5.1 Statistical data

Additionally to the parameter estimation, a posteriori site variance-covariance matrices and interpoint variance-covariance matrices should be produced.

11.5.2 Loop closures

Computing loop misclosures is the most direct method of performing quality control checks. The best loop software allows the user to choose arbitrary loops by specifying various routes by node identifier.

11.5.3 Residuals

Observation residuals resulting from the baseline adjustment can also be used for checking the homogeneity of the data. A plot of these residuals quickly shows outliers, and critical measurements can be isolated or eliminated.

11.5.4 Repaired cycle slips

A listing of the repaired cycle slips can also be helpful for understanding the quality of the data.

11.6 Network adjustment

11.6.1 Helmert transformation

The final results of a baseline vector computation are the differences in ECEF coordinates. The coordinates of points are then determined by combining these coordinate differences in a network adjustment where one (or more) point is held fixed. A seven parameter similarity transformation (i.e., Helmert transformation) is generally performed in order to shift, scale, and rotate the entire network to fixed control. Appropriate software is an essential part of GPS surveying.

11.6.2 Hybrid data combination

More flexible network adjustment programs that adjust both GPS vectors and conventionally measured angles, distances, and elevation differences are useful when GPS and conventional measurements are combined. The combination of GPS results with any other kind of data is optimally performed by the least squares collocation method.

11.7 Data base management

11.7.1 Storage and retrieval of data

The measured data of GPS surveys should be organized in a data base. This permits retrieval of the data for additional computation. For example, recomputation using precise orbits instead of broadcast orbits or using a changed pattern of the fiducial stations should be possible.

11.7.2 Archiving survey results

The results of the processing should be archived by means of the data base. This ensures an easy access to the data for any purpose. File compaction programs can be used to good advantage in archiving data.

11.7.3 Interface to national control points

A plot of both the national network reference points and the project points is useful in determining which points to connect by measurements. Generally, new project points closest to the national control points should be selected as tie points.

11.7.4 Interface to GIS software

Including geodetic data as a Geographic Information System (GIS) overlay or CAD file permits the viewing of point locations and provides users with a clear understanding of actual site locations (i.e., their juxtaposition with streets, rivers, contour lines, etc.).

11.8 Utilities

11.8.1 File editing

Occasionally, editing of a data file is necessary. For example, data from noisy epochs might be eliminated, data from fragmented files might be combined, or large files might be split into smaller files.

11.8.2 Time conversions

In addition to GPS time, there are other time systems desirable such as the Gregorian date, the modified Julian date, and others.

11.8.3 Optimization of survey design

There are various designs for a survey. In some cases maximum accuracy is desired, while in other cases minimum cost is the objective.

11.8.4 Transformation of coordinates

The coordinates obtained as a result of the baseline adjustment are given in the WGS-84 coordinate system as X, Y, Z coordinates. Transformations of these ECEF coordinates to latitude, longitude, and height should be included in the software suite. Transformations to frequently used coordinate systems (e.g., NAD-83, NAD-27, state plane) should also be included. More generally, transformations from a global datum to a local datum are necessary. Software should also be included to map geodetic coordinates to a map projection (e.g., UTM, Lambert projection, Gauss-Krüger projection).

11.8.5   Documentation of results

The final project documentation should contain a complete list of the results. Various plots (e.g., network points and baselines, residuals) are helpful to understand the survey results.

11.9   Flexibility

This section partly repeats items previously mentioned and stresses some additional aspects which may help evaluate the versatility and flexibility of a particular software package.

- Computer compatibility. There are some minimum computer requirements; however, software should run on a wide variety of computers. As far as possible, the software should be computer independent.

- Processing time (computer dependent). Computation time is of course a crucial question and is strongly dependent on the complexity of the software. Using a single baseline as a test criterion is a good method of timing various programs.

- Memory amount.

- Including additional data. It should be possible to add one or more data sets at a point to create a larger single data set. For example, combining two sessions at a point.

- Limitations. What are the limits for the number of sites, satellites, sessions, measured data?

- Data. An option for reformatting the observation data into the RINEX format is essential. It should also be possible to process both single and dual frequency data using the same software. Combining data from different receivers (also for a single baseline) and possibly of different satellite types is highly desirable. Software should be capable of processing several sessions in either batch or manual mode.

- Orbits. It is critical that the software is capable of using both the broadcast and precise ephemerides. The possibility of computing the six Keplerian elements and their variations with time is a feature that is useful for both research and many practical purposes.

- Option to process single points, single baseline vectors, networks.

- Real-time applications. Software/hardware combinations that are capable of operating in real-time are becoming more important. These applications include real-time differential navigation and "stake-out" surveying. Communication links are critical to this type of GPS use.

- Parameter estimation. Is it possible to select the number of parameters for the vector computation or the adjustments or is it restricted to fixed parameters?

- Transformations and map projections. Are coordinate transformations widely offered? Transformations from the global system to a local datum are necessary. Map projections should also be available.

- User interactivity. There should be an option to change the amount of (necessary) user interactivity. For normal processing the user does not need or want much interactivity.

- User friendliness. How much time is necessary to learn how to use the software? Does it need an expert with lots of computer experience? Does it offer help routines?

11.10   Checklist for software modules

This section is a summary of the previous sections of Chap. 11. The most pronounced features for each of the described software modules are presented in tables. The goal of these tables is twofold. First, they can be used as a kind of checklist when testing the capabilities of a software package and, second, they can be a support for the development of a new software. Only keywords are given to keep the tables as concise and compact as possible. The tables treat the following topics:

- Module on planning
- Module on data transfer
- Module on data processing
- Module on quality control
- Module on network adjustment
- Module on data base management
- Module on utilities
- Module on flexibility

The tables are numbered $11.2, 11.3, \ldots, 11.9$ in agreement with the corresponding section numbers.

Table 11.2. Module on planning

Planning
• Satellite visibility
- rise/set times
- elevation, azimuth
- site obstruction
- plot
• Satellite geometry
- DOP
- plot
• Simulations
- observation period
- cofactor matrix
• Receiver preprogramming
- survey start/stop times
- multiple sessions

Table 11.3. Module on data transfer

Data transfer
• Downloading data
- hard disk
- cassette
- floppy
- cable
- radio link
• Decoding data
- navigation message
- measured data
- additional information
• Checking transferred data
- consistency
- integrity
• File handling
- site names
- fiducial points
- sessions
• Baseline definition
- all combinations
- independent baselines

Table 11.4. Module on data processing

Data processing
• Generating RINEX formatted data
• Ephemerides
- convert broadcast to orbit file
- input of precise ephemerides
• Code data processing
- single point solution
* position
* clock
• Phase data processing
- undifferenced phase
- phase differences
* single-difference
* double-difference
* triple-difference
- frequency combinations
* single frequency
* dual frequency
* ionospheric-free combination
* others
- phase/code combinations
• Data analysis
- gross error detection
- cycle slip detection and repair
- plots
• Covariance matrices
• Modeling the atmosphere
- ionosphere
- troposphere
• Parameter estimation
- computation of coordinates (baseline vectors)
* static surveys
* kinematic surveys
* single baseline mode
* multipoint mode
* combination of various sessions
- clock parameters
* offset
* drift
- ambiguities
* estimating (real values)
* fixing (integers)
- orbit
* orbital elements
* partial derivatives of orbital elements

Table 11.5. Module on quality control

Quality control
• Statistical data
- a posteriori site covariance matrices
- interpoint covariance matrices
- plots
• Loop misclosures
• Residuals
• Repaired cycle slips

Table 11.6. Module on network adjustment

Network adjustment
• Helmert transformation
• Hybrid data combination

Table 11.7. Module on data base management

Data base management
• Storage and retrieval of data
• Archiving survey results
• Interface to national control points
• Interface to GIS software

Table 11.8. Module on utilities

Utilities
• File editing
• Time conversions
• Optimization of survey design
• Transformation of coordinates
• Documentation of results

Table 11.9. Module on flexibility

Flexibility
• Computer compatibility
• Processing time (computer dependent)
• Memory amount
• Including additional data
• Limitations on the number of
- sites
- satellites
- sessions
• Data
- RINEX
- combination of single and dual frequency data
- combination of various sessions
- combination of different receiver types
- combination of different satellite systems
• Orbits
- broadcast ephemerides
- precise ephemerides
- modeling of orbital elements
• Options on computing
- single point
- single baseline
- network
• Real-time applications
- navigation
- relative positioning
- differential GPS
- communication link
• Parameter estimation
- selection of parameters
• Transformations and map projections
• User interactivity option
- high
- low
• User friendliness
- help routines

12. Applications of GPS

12.1 General uses of GPS

The general overall uses of GPS are numerous. In this chapter an arbitrary grouping of global, regional, and local uses has been selected. A further additional arbitrary classification of navigation and survey uses is made to help organize the discussion of GPS applications.

Before mentioning applications in detail, an overview of the present achvievable accuracies in static and kinematic surveys is given in Table 12.1. The partitioning of this table is performed by the two groups of static and kinematic surveys. The accuracies shown are based on the baseline length, the number of satellites, and the session duration.

Table 12.1. Accuracies for relative positioning

Frequency	Baseline [km]	Satellites	Session [min]	Accuracy [ppm]
STATIC SURVEYS				
Single	1	4	30	5–10
		5	15	
	5	4	60	5
		5	30	
	10	4	90	4
		5	60	
	30	4	120	3
		5	90	
Dual (codeless)	100	5	120	0.1
Dual (P-code)	50	4	10	1.0
	100	5	60	0.1
	500	5	120	0.1–0.01
KINEMATIC SURVEYS				
Single	3	5	0.1	10
			3	3
Dual (P-code)	10	5	0.1	3

The accuracies of Table 12.1 can be considered as a kind of average because better and worse results can be achieved under certain circumstances. Note that for example a fairly reasonable DOP value is presupposed.

12.1.1 Global uses

Navigation. This was the planned primary use of GPS. Both military and civilian uses of the system in this mode are similar in that users wish to know their spatial locations as precisely as possible. For example, all types of aircraft and vessels will use GPS for en route navigation and the denial of accuracy does not materially affect this use. The global 100 m level of accuracy achievable even with activated SA is not a problem for en route positioning. But as aircraft are ready for landing or as vessels enter restricted waters, accuracy becomes more critical, cf. Table 12.2. Another factor critical to aircraft navigation is the high level of integrity demanded; for example, by the Federal Aviation Administration (FAA). This agency has specified a status warning time of 30 seconds for en route navigation and 10 seconds during approach and landing. This level of integrity is achieved by either Receiver Autonomous Integrity Monitoring (RAIM) which is an internal quality control of the actual GPS solution, or by integrity channels which use signals from a source external to GPS, cf. Montgomery (1991).

Surveying. Global application of GPS is a powerful tool for geodesy. This science is involved in monitoring global geodynamical phenomena such as earth rotation or plate tectonics. Until now, Very Long Baseline Interferometry (VLBI) and Satellite Laser Ranging (SLR) techniques have been used for this purpose. GPS is not presently capable of replacing these techniques but it will be used to augment them and provide more cost effective solutions to geodetic problems.

Table 12.2. Current air user requirements in meters after Lachapelle (1991)

Phase	Category	Position	Height
En route		≥ 100	≥ 100
Approach	Cat. I	17.1	4.1
and	Cat. II	5.2	1.7
landing	Cat. III	4.0	0.6

Timing and communications. Another global use of GPS is in global determination of accurate time. High accuracy timing has many scientific applications such as coordinating seismic monitoring and other global geophysical measurements. Inexpensive GPS receivers operating on known stations provide a timing accuracy of about 0.1 microsecond with only one satellite in view. With more sophisticated techniques, one can globally synchronize clocks even more precisely. Presently, the achievable accuracy of time transfer via GPS is some tens of nanoseconds, however, one nanosecond is considered possible.

Global communication depends upon precise coordination to pack as many data bits into a given period of time as possible. GPS should make it possible to increase the overall efficiency of communication, allowing many more users per unit of time than is presently possible.

12.1.2   Regional uses

Navigation. The regional applications of GPS navigation are enormous and include: exploration, transportation management, structural monitoring, and various types of automation.

The denial of full system accuracy can be overcome by differential navigation. This involves placing fixed GPS receivers at known locations in a project area and broadcasting range corrections to receivers within the area. The differential pseudoranging technique is also being used for hydrographic surveys. The U.S. Coast Guard is in the process of testing differential navigation networks to assist vessels approaching the coast. Additionally, private firms have set up similar networks in heavily trafficked areas and sell a precise navigation service. These networks consist of GPS receivers strategically located at known points where all satellites are tracked on a continual bases. Range corrections are computed by subtracting the theoretically correct range to a satellite from the measured range. The corrections for all satellites are then broadcast either by radio or satellite link to users who apply the correction to their measured satellite range to obtain a more accurate range. This corrected range is then used to compute the vessels' position. Differential navigation can virtually eliminate the effect of denial of accuracy.

There are numerous uses of GPS in managing regional resource and controlling activities within large areas. One example is the Geographic Information System (GIS) which is a computer based geographic oriented data base that helps manage resources on both regional and local levels. The GIS also provides a means of making rapid informed decisions for planning, development and tracking of infrastructure. Normally, a GIS is initialized by performing base mapping to create a digital data file that contains all map information. The digital file usually contains roads, buildings, vegetation types, property lines, soil types, and other critical data which can be displayed in various overlays on a computer screen. GPS often plays a key role in providing ground control for the base mapping. The coordinates of photo-identifiable points are determined by GPS and these points are used to control the scale and orientation of the map. GPS can also be used to update GIS information and enter new information in the file. This is accomplished by placing the GPS receiver over the point one wishes to include in the GIS to determine its location and then keying in an object code for the item so that it is tagged with time and coordinates.

The GIS is an excellent example of the use of GPS in its various modes: (1) point positioning, (2) precise surveying, (3) differential positioning, and (4) combinations of the methods.

The point positioning mode would be used to help locate existing survey marks whose coordinates had been previously determined, or whose coordinates had been scaled from a large-scale map. Inexpensive hand held receivers are being used to locate control points that have defied recovery because land features have greatly changed. Also, this method facilitates locating property corners whose coordinates have been computed from surveyed bearings and distances from a known point.

Next, the survey mode is used to measure the precise vectors between the known control points (both horizontal and vertical) and selected photo-targets or photo-identifiable points. These points are used in the photogrammetric map compilation to scale and orient the photograph to the chosen datum. The static, pseudokinematic, and kinematic modes could all be used during this phase of the work. Recently, the kinematic method has also been used to determine the coordinates of the photocenter by placing one GPS receiver in the aircraft and a second on a ground datum point. The P-code receivers show particular promise for this application since they will be able to resolve cycle slips or interruptions to the received signal during flight. This is accomplished by using the smoothed P-code pseudorange to compute the vector between fixed and airborne receiver using the differential navigation technique. The integer bias of the wide lane phase and, finally, the integer ambiguity of the base (19 cm) carrier is resolved to reinitialize the airborne receiver on-the-fly. This technique significantly reduces the amount of ground control needed to produce a metric map.

After the map has been produced and the GIS data base initialized, changes and corrections often must be made to update the data base. It would be inordinately expensive to refly and remap the area, so that GPS can further be used to determine the coordinates of additional points to be placed in the data base. For example, instrumented vans are being used to collect information as they drive along the highway. Videocameras and other devices collect data while GPS keeps track of the position of the van, and thus the various features are collected. The positional accuracy of the roving GPS receiver can be improved by use of the differential navigation mode. Several state highway departments are using this technique to develop a data base of all important road features so they can better manage their resources.

Surveying. Geodesists have long desired to measure crustal movements for various scientific purposes. One use would be to predict earthquakes by measuring certain precursor ground movement. GPS is an ideal tool for such studies in that the equipment is relatively inexpensive, portable, and highly accurate. For example, the NGS has measured an array of widely spaced points in the eastern part of U.S. (Eastern Strain Network) using dual frequency GPS receivers. This array or network will be periodically resurveyed so that crustal movement can be measured. Typically, this information can be used to provide nuclear plant siting information as well as providing a general purpose high accuracy network. The NGS network is tied into or referenced to GPS receivers located at VLBI tracking sites so the Eastern Strain Network also provides a multipurpose high precision network. Various other groups have measured similar networks to determine both continental and island crustal motion throughout the world. For example, the Japanese have established an array of permanent GPS tracking stations to provide near real-time crustal motion monitoring. Another example is the geodynamical goal of the Austrian GPS Reference network (AGREF) presented by Stangl et al. (1991).

GPS is the first measuring device that can be used to accurately measure height differences in real-time. Studies by the NGS have shown that repeat measurements between stable and subsiding points provide accurate measures of the subsidence. Industry has used GPS to measure the subsidence among groups of offshore oil platforms by making repeat surveys. One example is the subsidence measurement of several North Sea oil platforms with respect to other platforms that were thought to be in a more stable area. Monthly GPS surveys showed measurable height changes, and trends in the results positively proved which platforms were subsiding (relative to other platforms). In this particular case, GPS was probably the only method to make the determination.

As GPS equipment decreases in price, it is expected that receivers will be permanently mounted on structures and at selected sites and used to measure both subsidence and crustal deformation.

12.1.3   Local uses

The difference between regional and local uses of GPS is quite arbitrary because a small state GIS could either be considered regional or local in scope. Also, the difference between navigation and surveying will decrease as navigation becomes more accurate. Nevertheless, the grouping is kept for convenience.

Navigation. As stated earlier, the air user accuracy requirements are fairly high during approach and landing, cf. Table 12.2. In order to achieve these accuracies in the navigation mode, fixed receivers could broadcast range corrections to incoming aircraft so that they can compute more accurate positions as they approach the runway. GPS aided approach and landing may be an economical answer to airport control in the future.

The use of GPS in its point positioning mode is becoming popular for emergency vehicle management. For example, all emergency vehicles are equipped with a C/A-code pseudoranging receiver and a radio transceiver. The position of each vehicle is determined by GPS as it travels to a designated area and the transceiver sends this position to a central dispatcher. The location of all vehicles can then be displayed on a screen so that at any given instant the dispatcher can see where his resources are positioned. If, for example, one of the roads were blocked, the dispatcher could reroute the vehicle or he could dispatch a closer vehicle. There is little doubt such GPS based Automatic Vehicle Location (AVL) systems will be installed in all major cities in the world in the near future.

Surveying. The GIS discussed in the previous section is also being developed for cities and towns so this use will not be covered further. A typical local use of GPS is local property and site survey. Some surveyors are using GPS to place all projects surveyed on a single datum (e.g., NAD-83). An economical way to do this is to place an antenna at a fixed central location (e.g., office) and to determine the precise coordinates of this point by measuring vectors from the nearby fixed control points. After the coordinates of the fixed central site have been determined, one receiver is left there and is run continuously during times field crews are engaged in survey activities. Each survey crew takes one or two additional receivers and places them on points in the local scheme being surveyed. For example, a topographic site survey and boundary determination requires the establishment of many temporary and permanent points all connected by conventional survey measurements. Two GPS receivers placed on two widely spaced points in this local scheme could collect data during the time the survey crew made conventional measurements with little additional manpower cost. Upon completion of the conventional survey and simultaneous GPS measurements, the coordinates of all points in the scheme would be determined and also all bearings in the scheme would be precisely related to true north.

An additional local use of GPS is to perform topographic surveys by using the kinematic mode. In areas of relatively few obstructions, a roving GPS receiver can either be carried or placed on a vehicle (e.g., all terrain vehicle) and the terrain traversed by a series of cross sections. The horizontal position and the height of points can typically be determined every second, so a high density of point determinations will result even when the receiver is vehicle mounted and travels at high speed. The processing and plotting of the kinematic cross sections is automated so that the field to finish time can be minimized.

Several organizations have used GPS kinematic surveying to determine the coordinates of the photocenter during aerial mapping flights. When the GPS receiver is carefully synchronized with the mapping camera, accuracies of a few centimeters have been obtained. There is also consideration being given to use three or more antennas connected to the GPS receiver to determine the three orientation angles of the aircraft as well as its position. In any case, the knowledge of an accurate air base significantly reduces the ground control that will be needed to perform even large-scale mapping.

12.2 Installation of control networks

GPS is in fact a geodetic tool in that it provides precise vector measurements over long as well as short distances. Virtually all GPS processing software employs a three-dimensional model and results are given in geodetic latitude, longitude, and ellipsoidal height in the WGS-84 system which for vectors is (virtually) the same as NAD-83. If the GPS datum does not correspond to the national datum, a 3D similarity transformation must be performed. Afterwards, the ellipsoidal values are mapped onto a projection so that land surveyors who are more familiar with plane coordinates can use the survey results.

GPS provides reference datums by two methods: (1) the passive control networks, and (2) the active control networks.

12.2.1 Passive control networks

Virtually all civilized areas have some type of geodetic control networks generally surveyed by triangulation, traverse (or a combination of the two methods), and by spirit leveling. In many cases, the horizontal control network is separate from the vertical control network with perhaps 10% of these points in common. These control networks were used to scale and orient maps to provide a unified reference frame for large-scale projects, and to provide a common datum for property surveys. Geodetic control networks basically serve two functions: (1) they provide a "seamless" datum spanning large areas, and (2) they provide a reference framework which appears to be errorless to its users (surveyors).

The apparent errorless aspect of a geodetic network is accomplished by employing state-of-the-art instrumentation and using the utmost care in making observations. Historically, geodetic survey instruments were manufactured by specialists, and observers were rigorously trained in their use. Horizontal accuracies of 1:250 000 of the distance and vertical accuracies on a few millimeters per kilometer were typically obtained.

The advent of GPS changed the concept of what constituted a geodetic control network. The first surveys performed in 1983 were carried out with internal accuracies approaching 1 ppm, and disagreed with the existing control networks by 10 to 20 ppm, in many cases. This discrepancy was not surprising to the geodesists responsible for establishing the control networks, but it did concern GPS surveyors because they could not obtain accurate position checks of their work. Eventually, the concept of "supernets" or sparse arrays of high accuracy control networks established by GPS began to evolve. Many such supernets are now being established in the U.S. and in other countries by state governments. These supernets will supersede the traditional lower accuracy control network.

Horizontal control. In addition to being less accurate than required by the users, the existing control networks have another disturbing feature that dooms them to extinction. The points in the 2D network are normally located on the highest point in the area and are often not easily occupiable by GPS. Also, existing control points are located in heavily treed areas or at sites that are difficult to reach. Since all future geodetic (horizontal) surveys will be performed by GPS, the existing network points are inadequate.

The solution to this problem was found by establishing an array of new control points spaced 25 to 100 km at locations suitable for GPS occupation. In most cases, these points are located at sites that are vehicle accessible and where they will not be disturbed. The internal accuracy of the supernets is normally 0.1 ppm so that in effect they appear to be errorless to users.

The global accuracy of supernets is controlled by referencing them to VLBI points. The international VLBI framework of points provides the ultimate geodetic framework, inasmuch as the interrelationship of these points is known to a higher accuracy than any other global array of points. The supernet observations using dual frequency receivers are made simultaneously with observations at the VLBI sites. In fact, these observations are routinely made by the CIGNET stations, cf. Sect. 4.4.1. Since the positions of the points in a state supernet are known to a few centimeters accuracy with respect to the VLBI stations, it is possible to establish networks in adjoining states at separate times. The seamless feature required for a true geodetic network is, therefore, easily achieved because there is relatively little error in the common boundary between the adjacent states. This feature of GPS supernets is particularly important because the surveys can be performed as needed and do not have to proceed in an orderly progression as conventional surveys.

Vertical control. GPS is a three-dimensional survey system in that it measures a vector between points with roughly the same degree of accuracy in all three components. Therefore, GPS is able to determine elevations as well as latitude and longitude. However, there is a problem in using GPS to establish elevations, since the height differences measured with GPS are referenced to an ellipsoid and not to the geoid to which conventionally leveled heights refer. As explained in more detail in Sect. 10.2.3, the difference between the ellipsoid and geoid differs for every point on the earth and can be found from existing global or local geoid models. One way to model the local or regional geoid is to run levels to a point of known ellipsoidal height. This is normally done for points in a supernet so that each point has an accurate ellipsoidal height and elevation. Based on this, geodesists are able to refine their geoidal height prediction model to provide a more accurate determination. Many geodesists feel that prediction models in the future generally will have accuracies of a few centimeters in areas where supernets have been established.

Vertical accuracies of a few centimeters can also be presently achieved by connecting or referencing a GPS survey to at least four supernet points. Since the supernet points have well-known elevations, and interpolation is done between these known points, the elevations of the new points can be accurately determined. Additionally, accuracies of such surveys are improved when a good model of geoidal undulations (i.e., difference between geoid and ellipsoid) is used.

12.2.2 Active control networks

Just as the supernets are making the existing control networks obsolete, the development of active control networks will possibly make the supernets obsolete. For example, the Canadian Active Control System (ACS) has been developed since 1985. When completely deployed, it will consist of about twenty regularly distributed monitor stations which continuously track all visible satellites. To date, four stations with the Master Control Station (MCS) in Ottawa are in operation. The operation is fully automated. Thus, satellite tracking, collection of dual frequency data, collection of ground meteorological data, performance of self-diagnostic tests, and the communication with the MCS in full duplex mode is performed automatically. The MCS coordinates the observing stations, processes the data, and manages the dissemination of the results. For more details the reader is referred to Delikaraoglou et al. (1990).

Another prototype of an active control network is the system developed by the Texas Highway Department. The plan is to continuously track all satellites from nine sites located throughout the state and then use the phase data to compute vectors from these fixed sites to receivers surveying the various highway projects. The Texas system is a good example of the advantages and disadvantages of this concept. The advantage is that survey crews do not have to search for and occupy control points near their project area. They simply occupy the points they need coordinates for at the given project site. The vectors from the fixed stations surrounding the project site are combined with the (short) interstation vectors to give the "best fit" solution for that particular survey. The cost effectiveness of this approach is positive when numerous crews take advantage of the system so that the cost of maintaining the fixed sites is spread over a large number of units.

The disadvantage of the Texas system (at this time) is that the highway department is not prepared to provide its internal service to the public, as it is not funded to perform this service. The highway department's active control network is a single user private network which may have to be duplicated by various other users. This is not the only example of such duplication. Several other organizations are also setting up various types of active control points to service their particular needs. Only time will tell how this problem will be resolved.

In practice, any moderately sized organization (e.g., company) will be able to establish its own tracking network. For example, a firm could place three receivers at equally spaced points that would cover its operations area. The precise coordinates of these fixed sites could be determined by combining a day's observation with data from the CIGNET stations. Once the coordinates of the local tracking stations have been determined, they can be used as tracking stations themselves. The local tracking station can be as simple as an antenna mounted on the roof and a GPS receiver placed in a drawer and connected to a modem. At the end of the survey day, users would call each local tracker and download desired files. Recalibration of local networks could be performed periodically by recomputing the vectors from the CIGNET stations. The use of local tracking networks could be of significant advantage if the accuracy of the broadcast ephemerides is significantly degraded by SA. Local surveys would be unaffected by this degradation because surveyed vectors could be computed by using orbit relaxation software that only uses the broadcast orbit to initialize the computations.

12.3 Interoperability of GPS

12.3.1 GPS and Inertial Navigation Systems

One of the main disadvantages of using GPS for surveying, land navigation, and feature coordinate tagging is the temporary outage as the receiver passes under bridges or other obstructions. For example, the highway data collection van previously described would experience loss of positioning when the van passed under trees overhanging the highway. Various devices have been used to compensate for this disadvantage. One simple dead reckoning system to use is an odometer and a (flux gate) magnetic compass to assist in interpolating the vehicles' position during satellite outages, cf. Byman and Koskelo (1991). More sophisticated approaches involve combining GPS with Inertial Navigation Systems (INS).

An INS basically consists of two components: (1) gyros to stabilize space or locally oriented axes, and (2) accelerometers placed on them to measure accelerations (i.e., velocity rates). Hence, when mounting such an INS on a moving vehicle, one can continuously measure the accelerations along the stabilized axes. Starting from a known position and integrating the observables twice over time yields differences in position by which the trajectory of the vehicle is determined. This simple principle becomes fairly complicated in practice and the reader is referred to the voluminous literature, see e.g. Schwarz and Lachapelle (1990).

The error of an (unaided) INS generally increases with the square of time due to the double integration over time. What are the advantages of INS compared to GPS? First, inertial systems are autonomous and independent of external sources. Also, there is no visibility problem as with GPS. Secondly, INS provides similar accuracy to GPS when used over short time intervals with data rates of generally 64 Hz. Hence, INS can serve as an interpolator for GPS gaps.

When GPS is combined with a precise inertial navigation system, it is possible to reinitialize the GPS integer cycle count after a temporary loss of lock. For example, it is critical to maintain continuous count of integer cycles when performing a kinematic survey with C/A-code receivers. Accurate inertial systems could overcome the loss of cycle count problem by being able to predict an accurate change in carrier phase and thus cycle count during signal outages or massive cycle slips. The difficulty in using inertial systems for this purpose is their high cost and lack of field worthiness. Many experts feel that inertial systems become redundant for surveying if the P-code is left unencrypted for reasonable periods.

12.3.2 GPS and GLONASS

The GLONASS is the Russian equivalent of GPS. The nearly circular orbits with altitudes of 19 100 km and a period of 11.25 hours are similar to GPS. The system is expected to be completely deployed in 1995 and will be composed of 24 satellites (including three active spares) equally spaced in three orbital planes with 64.8 degree nominal inclination. In contrast to GPS, the broadcast carrier frequencies are satellite-specific. The individual carrier frequencies f_{L1}^j and f_{L2}^j for satellite j are defined by

$$
\begin{aligned}
f_{L1}^j &= 1602.0000\,\text{MHz} + j \cdot 0.5625\,\text{MHz} \\
f_{L2}^j &= 1246.0000\,\text{MHz} + j \cdot 0.4375\,\text{MHz}
\end{aligned}
\tag{12.1}
$$

where $j = 1, \ldots, 24$, cf. Kleusberg (1990b). The C/A- and P-code modulated onto the carriers are the same for all satellites and the code frequencies are about half the corresponding GPS values. Additionally, GLONASS is using different coordinate and time systems. Several manufacturers are planning to produce receivers that have dual GPS-GLONASS capability. The utility of such receivers for navigation as the GPS becomes operational is presently questionable although the obstacles may be surmountable. However, the added satellites would be of significant use in surveying applications. In a normal survey environment there are many obstructions that block satellite signals in certain portions of the sky. The addition of GLONASS satellites could significantly improve the possibility that a given line could be surveyed. There is much to be determined before all aspects of combined GPS-GLONASS systems are known.

12.3.3 GPS and other sensors

Experiments are also being proposed to place GPS receivers on low altitude satellites with various other sensing devices such as laser altimeters and synthetic aperture radar. Inclusion of GPS to determine positions of various events logged by such equipment will add to the data's value and in most cases reduce the cost of data collection.

One special use is to include GPS capability in the various earth sensing satellites. Accurate positioning of such satellites will greatly reduce the cost of determining geodetic coordinates of points appearing in the various images.

12.3.4 GPS and terrestrial survey

Several manufacturers also plan to combine GPS and conventional survey instruments. Today, the state-of-the-art surveying instruments measure angles and distances electronically so that the addition of GPS capability fits well with the total survey concept. Such a system would be particularly useful once a network of active control points were established because continuous accurate positioning would then be possible by computing vectors to the surrounding active control points.

13. Future of GPS

13.1 New application aspects

The future uses of GPS are limited only by one's imagination. Many of
the present uses were described in various articles written as early as 1982
when this new system first demonstrated its high accuracy. With the reduction of equipment costs and the increase in satellite coverage many more
applications will develop.

One such application which has been mentioned is the use of GPS to
automate various types of machinery. For example, it should be possible to
automate the grading and paving equipment used for road building. Equipment could be run around the clock without operators with GPS performing all motion operations based upon a digital terrain model stored in the
equipment's computer. Similarly, ships and planes could be operated in an
automated mode with takeoffs and landings being performed by integrated
GPS/computer units. The remainder of this chapter will deal, primarily,
with the future use of GPS in surveying related activities.

13.2 Impact of limited accuracy and access

A description of selective availability (SA) and anti-spoofing (A-S) was already provided in Sect. 2.2.3. Hence, this discussion will focus on the effect
these two system limiters will have on surveyors.

13.2.1 Selective availability

The activation of SA did not significantly affect many surveying uses of
GPS. Denial of accurate real-time (broadcast) ephemerides by transmitting
a truncated navigation message primarily affects the navigation users of the
system. Geodetic surveyors can normally wait until after the measurements
have been completed to compute the desired coordinates. Land surveyors,
on the other hand, need the coordinates of selected points immediately (i.e.,
when laying out points), but can tolerate the errors introduced by less accurate ephemerides.

The lack of broadcast ephemerides accuracy can be overcome by the establishment of private tracking stations or active control points. In the case
of geodetic surveys where some days can be tolerated between the measurements and final computations, postprocessed ephemerides will suffice. In

cases where it is imperative that precise coordinates be determined immediately, the use of active control stations is more appropriate. Some manufacturers have already demonstrated the real-time broadcast of carrier phase data between a fixed and roving receiver so that the coordinates of the roving receiver can be determined instantaneously by the kinematic technique to at least centimeter accuracy, cf. Ferguson and Veatch (1990) or Euler et al. (1992). This mode could be refined and applied to active control points so that the precise coordinates could be determined in the field without reference to the official ephemerides and, consequently, would negate the effect of ephemerides denial. The error which results from imprecise ephemerides is quite different for point positioning and differential surveying (either positioning or surveying). In general, point positioning errors are of the same magnitude as the satellite (orbital) error. In the case of relative positioning, relative baseline errors correspond to relative orbital errors.

The second method of SA is also somewhat negated by the procedures used in surveying. The present dithering of the satellite clock signal produces range noise that translates into point positioning errors. This form of SA also primarily affects real-time navigation using a single receiver. When two receivers are used to measure the vector between them, as in the surveying mode, the dithering can be negated by ensuring that the receiver measurement time tags are accurate. Vector computations difference the phase between receivers so that, when the time tags are correctly made, any dithering offset is canceled. The dithering effect can also be negated in the navigation mode by the use of active control points that broadcast range correctors. The U.S. Coast Guard plans to measure and broadcast satellite range corrections from a number of fixed sites along the coast. Use of these corrections virtually eliminate the effects of SA. At present, the effects of SA are less than the 100 m two-dimensional position and 159 m height mean square error planned for the system. This will likely not change with full constellation except in severe military situations.

Many other governmental and private organizations have plans to set up active control points to measure both range corrections and phase data. In the future, it is hoped that these activities can be coordinated. The cost of measuring both the code range and carrier phase is not much higher than measuring range corrections so that active control stations established for navigation could also double as high-order survey control points.

13.2.2   Anti-spoofing

The other technique for denying full access to the GPS is to activate a code that makes it impossible to lock onto the P-code broadcast on both the

$L1$ and $L2$ carriers. Denial of the P-code limits the full exploitation of the system and prevents such uses as the wide lane technique for the resolution of integer ambiguities on-the-fly (for long lines). In the static surveying mode, the phase of the $L2$ carrier can be measured using the codeless technique and long lines can be accurately measured by reducing the ionospheric noise.

The decision to activate A-S on a full time basis will likely be affected by the final development of GLONASS and by global politics. It has been suggested by experts that the P-code denial could be periodically implemented for testing and that it would be invoked during times of war. If A-S is not implemented full time, P-code receivers could be used for surveying more cost effectively than C/A-code receivers. Because high signal strengths are available on both frequencies, accurate phase measurements can be made on $L1$ and $L2$. These measurements can be combined to produce the wide lane signal with a wavelength of about 86 cm. The integer ambiguity of this combined wavelength is more easily resolved than the shorter 19 cm base carrier wavelength. Experts feel that P-code receivers will be able to resolve the combined integer bias using the smoothed P-code pseudorange, and the various combinations of $L1/L2$ wavelength biases can be used to solve the integer bias of the corrected base carrier. Even a direct ambiguity resolution is possible in the event the accuracy of the pseudoranges drops below half the carrier wavelength. When all this works, roving pseudokinematic receivers will be able to be initialized on-the-fly so that cycle slips can be detected and repaired during movement of the receiver. This capability will result in virtual centimeter navigation, which has numerous applications including control of aerial photography and hydrographic surveys.

The second surveying use of P-code receivers is to perform rapid static surveys. A recent P-code receiver tested by the FGCC in 1991 demonstrated that full accuracy could be obtained with data of 10 minutes using the wide lane technique. This technique is similar to the method described above where the integer ambiguity of the combined wavelength is first determined to help resolve the $L1$ (19 cm) wavelength. The cost effectiveness of P-code receivers for static surveys will be significantly higher than C/A-code receivers because with a fixed manpower cost the productivity is greatly increased.

13.3 Improved constellation

In March 1992, there are five operational Block I satellites most of which are approaching the end of their design life. The twelve operational Block II satellites should still have several years of life in the mid 1990's when it is estimated that GPS will be fully operational so that problems in completing the constellation on time are not anticipated.

13.3.1 Next generation satellites

The Block IIR or follow-on satellites are presently being designed to include improvements over the present Block II satellites. It is planned to include the capability to transmit measurements between satellites and to transmit the results of these measurements between spacecraft. This capability will allow the satellites to essentially position themselves without extensive ground tracking. The advantages of these satellites are:

- Maintain navigation accuracy for six months without ground support. Survivable control not required. No user modifications required.

- Minimize uplink jamming concerns.

- One upload per spacecraft and month instead of one or even more per day.

- Reduce need for overseas stations to support navigation uploads.

- Improved navigation accuracy.

These improved features mainly benefit the military use of the system, since civilians will still be required to provide their own ephemerides for accurate surveys.

13.3.2 GLONASS satellites

In addition to the GPS constellation, the GLONASS constellation will possibly provide a like number of satellites which can be used in combination with the GPS satellites. Because of the similarity of the two systems, it will be possible to include the GLONASS measurements in a combined solution forming triple- and double-differences between the satellite pairs. The main benefit to surveyors will be the increase in sky coverage, making it possible to measure lines that would otherwise be obstructed. Also, the increased number of measurements makes the kinematic method much more viable than it is with just the GPS constellation. For example, Remondi (1991a) has shown that kinematic surveys can be performed without presurvey initialization when five or more satellites are available and using C/A-code receivers. By the integration of GLONASS kinematic surveys can proceed even after cycle slips have been experienced. On-the-fly kinematic surveying using an inexpensive C/A-code receiver could be used for most present land surveying tasks such as boundary, topographic, hydrographic, and construction surveys.

13.3.3 INMARSAT satellites

The INternational MARitim SATtellites, a system of geosynchronous communication satellites, are planned to be equipped with repeaters which will transmit C/A-code signals on $L1$, cf. Kinal and Singh (1990). The primary purpose of these satellites is integrity monitoring. However, these satellites will significantly increase the effective coverage of GPS and will be particularly useful in the kinematic surveys since satellite motion is not needed. Also, these satellites can be used as the reference satellite in static surveying. A detailed description of INMARSAT can be found in Ackroyd and Lorimer (1990), pp. 171–198.

13.4 Hardware improvements

Most of the hardware developments existing today were predicted by many of the GPS pioneers. Also, the rapid cost reduction in equipment has been predicted, cf. Wells et al. (1987), and in many ways parallels the development of small computers. The continued production of special computer chips will both decrease the cost and size, and increase the reliability and field worthiness.

13.4.1 Receiver cost

The first GPS receivers for commercial use cost in excess of \$ 150 000 and possessed much less capability than receivers costing \$ 20 000 today. The Original Equipment Manufacturer (OEM) units coming into the market in 1991 cost less than \$ 10 000 and claim to give survey accuracies. If one plots the receiver cost against time, the curve shows a sharp (exponential) decline during the first few years and then becomes asymptotic and slowly decreases in price as time increases. The GPS price curve closely resembles the price of survey totalstations or hand held calculators. The cost analysis of GPS is somewhat complicated by the fact that the processing software is bundled in the sales price. When the cost of software is a significant portion of the receiver cost, and the receiver is mainly composed of integrated circuit chips, the cost of receivers can be appreciably lowered by quantity production. Today, the point has been reached where volume production by a manufacturer could reduce the receiver cost to a fraction of the present level.

13.4.2 Receiver capability

The present multichannel P-code receivers measure and record all the basic observables; that is, pseudoranges with both codes and phases on both carriers. The resolution will be improved to 0.1% of the wavelength or even better. This means that a precision in the P-code pseudorange of a few centimeters is achievable which allows for instantaneous ambiguity resolution. Additionally, multichannel receivers with up to 36 digital programmable channels that can be programmed to observe any combination of the observables will be manufactured. For example, three or four antennas can be connected to a single receiver to determine position and attitude. It is anticipated that such receivers will be modified to determine azimuths to distant azimuth marks, thus, reducing by 50% the task of surveying a geodetic control network.

Improvements with regard to kinematic GPS will also affect the design of future GPS receivers. Automatic bandwidth adjustment and more rigorous cycle slip detection will be the norm of the future. As stated earlier, by the use of new technologies the size of receivers will also be reduced, especially if they are to be used in conventional topographic and construction surveys. It is easy to imagine a small antenna mounted on top of a range pole with the electronics perhaps placed inside the pole to reduce the bulk carried by the rodman.

GPS receivers will also be incorporated in conventional totalstation instruments. For example, one receiver in a totalstation and a second in the rod would significantly increase the surveyor's productivity. The totalstation could be used in those instances when the point desired was under a tree or otherwise obstructed.

As the cost of receivers decreases, instruments will also be placed on structures such as bridges and dams to provide near real-time motion detection. It has been hypothesized for a number of years that instrumenting dams with GPS receivers tied into a central control network could provide a warning system to foretell dam failure.

The GLONASS satellite system was briefly discussed previously. If this system develops as planned, receivers of the future will incorporate the capability to receive both GPS and GLONASS data. It is worth noting that prototypes of such receivers are already available. Receivers with programmable digital channels will be tailored to switch channels between the two systems to gain the optimum accuracy for the satellites in view. Also, the number of channels in receivers will increase if the GLONASS system is used since there could be as many as twenty useable satellites in view from the two systems.

13.5 Software improvements

There are three basic types of GPS software: (1) preplanning software, (2) the programs used by the receiver to track and record the satellite data, and (3) postprocessing software. Some GPS receivers presently include partial postprocessing software in the receiver. For example, one manufacturer offers the capability of performing real-time differential navigation using their normal hardware and upgraded receiver software. This trend will increase in the future and additional processing software will be included with the receiver. The prior generation TRANSIT receivers had the capability to load data from a second receiver into the master receiver to process the two data sets and determine more precise differential positions. It is likely that GPS receivers will also provide this capability in the near future. A similar development is receiver software that will notify the operator when a sufficient number of observations has been taken.

The first GPS processing software required that cycle slips be fixed by hand. A reasonably good line (few cycle slips) required about one hour to process on a DEC PDP-11 computer. A major breakthrough was made when C. Goad introduced processing software which automatically repaired cycle slips and also could be batched so that a day's data could be processed on an 8088 PC computer overnight. The kinematic and pseudokinematic software introduced by B. Remondi produced new and better ways to accurately measure vectors and added to the overall capabilities of the system. Today, some manufacturers are incorporating the wide lane technique to quickly determine integer biases and thus shorten the time required to measure vectors.

Following the general trend in receiver and constellation development, processing software will be designed to incorporate all combinations of the basic observables. Experts have postulated that improved P-code pseudoranges can be used to determine the wide lane bias which in turn can be used to determine the base (19 cm) wavelength biases. This capability will mean that on-the-fly kinematic initialization will be possible providing virtually centimeter accuracy navigation. If the GLONASS system becomes fully operational as planned, this system will be used to provide information that is missing or denied by the GPS so that full combined system capability will be maintained.

The CIGNET tracking network and active control points were discussed in previous chapters. In the future, these types of networks will be expanded and improved. Software will be developed so that surveyors can download data from one or more active control points in real- or near real-time using a mobile telephone. They will then be able to compute accurate positions on-site and to determine the degree of accuracy of the position. The orbit relaxation type software used for this purpose will provide centimeter accuracy positions over large areas so that the concept of survey accuracy will virtually disappear since all control points will be higher than first-order. In this scenario, the surveyor could place a second receiver on a nearby temporary point so that the azimuth between the two points can be computed at the same time as position.

13.6   Conclusion

Considering the present status of GPS and taking into account all the improvements previously discussed leads to an exciting outlook of the future of GPS. The enlarged number of next generation satellites will meet all the requirements of navigation, surveying, and timing. Substantially improved hardware and software components will provide the desired results more quickly or even in real-time. The real-time transfer of data will become cheap and easy in the near future. The accuracy is expected to be increased by at least one order of magnitude with lower costs. In addition, GPS will achieve a higher level of integrity and its output will be more reliable by extended quality control. As one consequence, the differences between navigation and surveying will be less pronounced. Apart from the present tasks, new application aspects arise at the horizon which nobody would have considered only a few years ago.

The future of GPS is really fascinating. However, what will happen within, let us say, the next decade can never be predicted exactly. Hence, instead of relying on general predictions one should carefully observe present developments in order to keep abreast of this rapidly changing technology.

References

Ackroyd N, Lorimer R (1990): Global navigation – a GPS user's guide. Lloyd's of London, London New York Hamburg Hong Kong.

Allison T, Eschenbach R (1989): Real-time cycle-slip fixing during kinematic surveys. In: Proceedings of the Fifth International Geodetic Symposium on Satellite Positioning, Las Cruces, New Mexico, March 13–17, vol 2: 330–337.

Arnold K (1970): Methoden der Satellitengeodäsie. Akademie, Berlin.

Arradonda-Perry J (1992): Receiver survey. GPS World, 3(1): 46–58.

Ashby N (1987): Relativistic effects in the Global Positioning System. In: Relativistic effects in geodesy, Proceedings of the International Association of Geodesy (IAG) Symposia of the XIX General Assembly of IUGG, Vancouver, Canada, August 10–22, vol 1: 41–50.

Ashjaee J (1986): GPS Doppler processing for precise positioning in dynamic applications. In: The Institute of Navigation: Global Positioning System, vol 3: 54–69.

Ashjaee J, Lorenz R, Sutherland R, Dutilloy J, Minazio J-B, Abtahi R, Eichner J-M, Kosmalska J, Helky R (1989): New GPS developments and Ashtech M-XII. In: Proceedings of ION GPS-89, The Second International Technical Meeting of the Satellite Division of The Institute of Navigation, Colorado Springs, Colorado, September 27–29, 195–198.

Ashkenazi V, Fuente C de la, Moore T, Yau J (1987): GPS processing algorithms and accuracies. Paper presented at the XIX General Assembly of the IUGG at Vancouver, Canada, August 10–22.

Ashkenazi V, Moore T, Ffoulkes-Jones G, Whalley S, Aquino M (1990): High precision GPS positioning by fiducial techniques. In: Bock Y, Leppard N (eds): Global Positioning System: an overview. Springer, New York Berlin Heidelberg London Paris Tokyo Hong Kong, 195–202 [Mueller II (ed): IAG Symposia Proceedings, vol 102].

Ashtech (1991): GPS field surveying techniques. Ashtech Incorporation, AN/GPS Field 5/91.

Avdis V, Billiris H, Hurst K, Kastens K, Paradissis D, Veis G (1990): Selection, monumentation, and documentation of GPS sites in the circum-Aegean region, Greece. Newsletter of the Space Geodetic Measurement Sites Subcommission (SGMS), 1(2).

Bastos L, Landau H (1988): Fixing cycle slips in dual-frequency kinematic GPS-application using Kalman filtering. Manuscripta geodaetica, 13: 249–256.

Bauer M (1989): Vermessung und Ortung mit Satelliten. Wichmann, Karlsruhe.

Bauersima I (1983): NAVSTAR/Global Positioning System (GPS) II, Radiointerferometrische Satellitenbeobachtungen. Mitteilungen der Satelliten-Beobachtungsstation Zimmerwald, Bern, vol 10.

Beck N, Duval JR, Taylor PT (1989): GPS processing methods: comparison with precise trilateration. Journal of Surveying Engineering, 115(2): 181–197.

Bender PL, Larden DR (1985): GPS carrier phase ambiguity resolution over long baselines. In: Proceedings of the First International Symposium on Precise Positioning with the Global Positioning System, Rockville, Maryland, April 15–19, vol 1: 357–361.

Bennet JM (1965): Triangular factors of modified matrices. Numerische Mathematik, 7: 217–221.

Beutler G (1991): Himmelsmechanik I. Mitteilungen der Satelliten-Beobachtungsstation Zimmerwald, Bern, vol 25.

Beutler G, Bauersima I, Gurtner W, Rothacher M (1987): Correlations between simultaneous GPS double difference carrier phase observations in the multistation mode: implementation considerations and first experiences. Manuscripta geodaetica, 12: 40–44.

Beutler G, Bauersima I, Gurtner W, Rothacher M, Schildknecht T (1988): Static positioning with the Global Positioning System (GPS): state of the art. In: Groten E, Strauß R (eds): GPS-techniques applied to geodesy and surveying. Springer, Berlin Heidelberg New York London Paris Tokyo, 363–380 [Bhattacharji S, Friedman GM, Neugebauer HJ, Seilacher A (eds): Lecture Notes in Earth Sciences, vol 19].

Beutler G, Bauersima I, Gurtner W, Rothacher M, Schildknecht T, Geiger A (1987): Atmospheric refraction and other important biases in GPS carrier phase observations. Mitteilungen der Satellitenbeobachtungs-station Zimmerwald, Bern, vol 22.

Beutler G, Bauersima I, Gurtner W, Rothacher M, Schildknecht T, Mader GL, Abell MD (1987): Evaluation of the 1984 Alaska Global Positioning System campaign with the Bernese GPS software. Journal of Geophysical Research, 92(B2): 1295–1303.

Beutler G, Davidson DA, Langley RB, Santerre R, Vanicek P, Wells DE (1984): Some theoretical and practical aspects of geodetic positioning using carrier phase difference observations of GPS satellites. University of New Brunswick, Canada, Technical Report vol 109.

Beutler G, Gurtner W, Bauersima I, Rothacher M (1986): Efficient computation of the inverse of the covariance matrix of simultaneous GPS carrier phase difference observations. Manuscripta geodaetica, 11: 249–255.

Beutler G, Gurtner W, Rothacher M, Wild U, Frei E (1990): Relative static positioning with the Global Positioning System: basic technical considerations. In: Bock Y, Leppard N (eds): Global Positioning System: an overview. Springer, New York Berlin Heidelberg London Paris Tokyo Hong Kong, 1–23 [Mueller II (ed): IAG Symposia Proceedings, vol 102].

Bevis M (1991): GPS Networks: the practical side. EOS Transactions, American Geophysical Union, 72(6), 49.

Black HD (1978): An easily implemented algorithm for the tropospheric range correction. Journal of Geophysical Research, 83(B4): 1825–1828.

Black HD, Eisner A (1984): Correcting satellite Doppler data for tropospheric effects. Journal of Geophysical Research, 89(D2), 2616-2626.

Blackwell EG (1986): Overview of differential GPS methods. In: The Institute of Navigation: Global Positioning System, vol 3: 89–100.

Bletzacker FR (1985): Reduction of multipath contamination in a geodetic GPS receiver. In: Proceedings of the First International Symposium on Precise Positioning with the Global Positioning System, Rockville, Maryland, April 15–19, vol 1: 413–422.

Blewitt G (1987): New approaches to GPS carrier phase ambiguity resolution. Paper presented at the XIX General Assembly of the IUGG at Vancouver, Canada, August 10–22.

Bock Y, Shimada S (1989): Continuously monitoring GPS networks for deformation measurements. In: Bock Y, Leppard N (eds): Global Positioning System: an overview. Springer, New York Berlin Heidelberg London Paris Tokyo Hong Kong, 40–56 [Mueller II (ed): IAG Symposia Proceedings, vol 102].

Boucher C, Altamimi Z (1989): The initial IERS terrestrial reference frame. Observatoire de Paris, International Earth Rotation Service, Technical Note 1.

Bowen R, Swanson PL, Winn FB, Rhodus NW, Feess WA (1986): Global Positioning System Operational Control System accuracies. In: The Institute of Navigation: Global Positioning System, vol 3: 241–257.

Bronstein IN, Semendjajew KA (1969): Taschenbuch der Mathematik. Teubner, Leipzig.

Brouwer D, Clemence GM (1961): Methods of celestial mechanics. Academic Press, New York.

Brunner F (1984): NAVSTAR GPS – ein Satellitensystem für die Navigation und Vermessung der Zukunft. Krieg im Äther, vol XXIV.

Bucerius H (1966): Himmelsmechanik I. Bibliographisches Institut, Mannheim, BI Hochschultaschenbücher vol 143/143a.

Byman P, Koskelo I (1991): Mapping Finnish roads with differential GPS and dead reckoning. GPS World, 2(2): 38–42.

Cannon ME (1990): The contribution of GPS to the information society. CISM Journal ACSGC, 44(3): 225–231.

Cannon ME, Schwarz KP (1989): GPS semi-kinematic positioning along a well-controlled traverse. In: Proceedings of the Fifth International Geodetic Symposium on Satellite Positioning, Las Cruces, New Mexico, March 13–17, vol 2: 621–630.

Cannon ME, Schwarz KP, Wong RVC (1986): Kinematic positioning with GPS – an analysis of road tests. In: Proceedings of the Fourth International Geodetic Symposium on Satellite Positioning, Austin, Texas, April 28 – May 2, vol 2: 1251–1268.

Chao CC (1972): A model for tropospheric calibration from daily surface and radiosonde balloon measurements. Jet Propulsion Laboratory, Pasadena, California, JPL Technical Memorandum, 391-350.

Chin M (1991): CIGNET report. GPS Bulletin, 4(2): 5–11.

Clynch JR, Coco DS (1986): Error characteristics of high quality geodetic GPS measurements: clocks, orbits, and propagation effects. In: Proceedings of the Fourth International Geodetic Symposium on Satellite Positioning, Austin, Texas, April 28 – May 2, vol 1: 539–556.

Collins J (1982): The satellite solution to surveying. Professional Surveyor, 2(6): 12–17.

Collins J (1989): Fundamentals of GPS baseline and height determinations. Journal of Surveying Engineering, American Society of Civil Engineers, 115(2): 223–235.

Counselman CC (1981): Miniature interferometer terminals for earth surveying – MITES. CSTG Bulletin vol 3, International Activities, Technology and Mission Developments.

Counselman CC, Abbot RI (1989): Method of resolving radio phase ambiguity in satellite orbit determination. Journal of Geophysical Research, 94(B6): 7058–7064.

Counselman CC, Gourevitch SA (1981): Miniature interferometer terminals for earth surveying: ambiguity and multipath with the Global Positioning System. IEEE Transactions on Geoscience and Remote Sensing, GE–19(4): 244–252.

Counselman CC, Hinteregger HF, Shapiro II (1972): Astronomical applications of differential interferometry. Science, 178: 607–608.

Counselman CC, Shapiro II (1978): Miniature interferometer terminals for earth surveying. Ohio State University, Department of Geodetic Sciences, Columbus, Ohio, vol 280: 65–85.

Counselman CC, Shapiro II, Greenspan RL, Cox DB (1979): Backpack VLBI terminal with subcentimeter capability. In: Proceedings of Radio Interferometry Techniques for Geodesy, NASA Conference Publication, vol 2115: 409–414.

Cross PA, Ahmad N (1988): Field validation of GPS phase measurements. In: Groten E, Strauß R (eds): GPS-techniques applied to geodesy and surveying. Springer, Berlin Heidelberg New York London Paris Tokyo, 349–360 [Bhattacharji S, Friedman GM, Neugebauer HJ, Seilacher A (eds): Lecture Notes in Earth Sciences, vol 19].

Davidson JM, Thornton CL, Vegos CJ, Young LE, Yunck TP (1985): The March 1985 demonstration of the fiducial network concept for GPS geodesy: a preliminary report. In: Proceedings of the First International Symposium on Precise Positioning with the Global Positioning System, Rockville, Maryland, April 15–19, vol 2: 603–611.

Davis JL, Herring TA, Shapiro II, Rogers AE, Elgered G (1985): Geodesy by radio interferometry: effects of atmospheric modeling errors on estimates of baseline lengths. Radio Science, 20(6): 1593–1607.

Decker BL (1986): World Geodetic System 1984. In: Proceedings of the Fourth International Geodetic Symposium on Satellite Positioning, Austin, Texas, April 28 – May 2, vol 1: 69–92.

Delikaraoglou D, Dragert H, Kouba J, Lochhead K, Popelar J (1990): The development of a Canadian GPS Active Control System: status of the current array. In: Proceedings of the Second International Symposium on Precise Positioning with the Global Positioning System, Ottawa, Canada, September 3–7, 190–202.

Delikaraoglou D, Lahaye F (1990): Optimization of GPS theory, techniques and operational systems: progress and prospects. In: Bock Y, Leppard N (eds): Global Positioning System: an overview. Springer, New York Berlin Heidelberg London Paris Tokyo Hong Kong, 218–239 [Mueller II (ed): IAG Symposia Proceedings, vol 102].

Dierendonck AJ van, Russell SS, Kopitzke ER, Birnbaum M (1980): The GPS navigation message. In: The Institute of Navigation: Global Positioning System, vol 1: 55–73.

Dong D, Bock Y (1989): Global Positioning System network analysis with phase ambiguity resolution applied to crustal deformation studies in California. Journal of Geophysical Research, 94(B4): 3949–3966.

Easton RL (1970): Navigation system using satellites and passive ranging techniques. U.S. Patent Office, Patent no. 3,789,409.

Egge D (1985): Zur sequentiellen Auswertung von Doppler-Satellitenbeobachtungen. Wissenschaftliche Arbeiten der Fachrichtung Vermessungswesen der Universität Hannover, vol 141.

Elgered G, Johansson J, Rönnäng B (1985): Methods to correct for the tropospheric delay in satellite-earth range measurements. In: Proceedings of the Second SATRAPE Meeting, Saint-Mandé, France, November 4–6, 60–76.

Essen L, Froome KD (1951): The refractive indices and dielectric constants of air and its principal constituents at 24 000 Mc/s. In: Proceedings of Physical Society, vol 64(B): 862–875.

Euler H-J, Hein GW, Landau H (1992): First experiences with a real-time differential GPS positioning system. Paper presented at the Sixth International Geodetic Symposium on Satellite Positioning at Columbus, Ohio, March 17–20.

Evans AG (1986): Comparison of GPS pseudorange and biased Doppler range measurements to demonstrate signal multipath effects. In: Proceedings of the Fourth International Geodetic Symposium on Satellite Positioning, Austin, Texas, April 28 – May 2, vol 1: 573–587.

Farrell WE (1972): Deformation of the earth by surface loads. Reviews of Geophysics and Space Physics, 10(3): 761–797.

Federal Geodetic Control Committee (FGCC) (1988): Geometric geodetic accuracy standards and specifications for using GPS relative positioning techniques, version 5.0.

Feissel M, McCarthy DD (1989): Explanatory supplement to IERS Bulletins A and B. IERS Information, 8.

Fell PJ (1980): Geodetic positioning using a Global Positioning System of satellites. Ohio State University, Department of Geodetic Sciences, Columbus, Ohio, vol 299.

Feltens J (1988): Several aspects of solar radiation pressure. In: Groten E, Strauß R (eds): GPS-techniques applied to geodesy and surveying. Springer, Berlin Heidelberg New York London Paris Tokyo, 487–502 [Bhattacharji S, Friedman GM, Neugebauer HJ, Seilacher A (eds): Lecture Notes in Earth Sciences, vol 19].

Ferguson KE, Veatch ER (1990): Centimeter level surveying in real-time. In: Proceedings of the Second International Symposium on Precise Positioning with the Global Positioning System, Ottawa, Canada, September 3–7, 1185–1195.

Finn A, Matthewman J (1989): A single frequency ionospheric refraction correction algorithm for TRANSIT and GPS. In: Proceedings of the Fifth International Geodetic Symposium on Satellite Positioning, Las Cruces, New Mexico, March 13–17, vol 2: 737–756.

Fliegel HF, Feess WA, Layton WC, Rhodus NW (1985): The GPS radiation force model. In: Proceedings of the First International Symposium on Precise Positioning with the Global Positioning System, Rockville, Maryland, April 15–19, vol 1: 113–119.

Frei E (1991): GPS-Fast Ambiguity Resolution Approach "FARA": theory and application. Presented paper at XX General Assembly of IUGG, IAG-Symposium GM 1/4, Vienna, August 11-24.

Frei E, Beutler G (1989): Some considerations concerning an adaptive, optimized technique to resolve the initial phase ambiguities for static and kinematic GPS surveying-techniques. In: Proceedings of the Fifth International Geodetic Symposium on Satellite Positioning, Las Cruces, New Mexico, March 13–17, vol 2: 671–686.

Frei E, Beutler G (1990): Rapid static positioning based on the fast ambiguity resolution approach: the alternative to kinematic positioning. In: Proceedings of the Second International Symposium on Precise Positioning with the Global Positioning System, Ottawa, Canada, September 3–7, 1196–1216.

Fricke W, Schwan H, Lederle T (1988): Fifth Fundamental Catalogue (FK5), Part I, The basic fundamental stars. Braun, Karlsruhe, Veröffentlichungen Astronomisches Rechen-Institut, Heidelberg, vol 32.

Geiger A (1988): Einfluss und Bestimmung der Variabilität des Phasenzentrums von GPS-Antennen. Eidgenössische Technische Hochschule Zürich, Institute of Geodesy and Photogrammetry, Mitteilungen vol 43.

Gervaise J, Mayoud M, Beutler G, Gurtner W (1985): Test of GPS on the CERN-LEP control network. In: Welsch WM, Lapine LA (eds): Proceedings of the Joint Meeting of FIG Study Groups 5B and 5C on Inertial, Doppler and GPS Measurements for National and Engineering Survey, Munich, July 1–3. Schriftenreihe der Universität der Bundeswehr München, vol 20-2: 337–358.

Gibson R (1983): A derivation of relativistic effects in satellite tracking. Naval Surface Weapons Center, Dahlgren, Virginia, Technical Report TR 83-55.

Goad CC (1985): Precise relative position determination using Global Positioning System carrier phase measurements in a nondifference mode.

In: Proceedings of the First International Symposium on Precise Positioning with the Global Positioning System, Rockville, Maryland, April 15–19, vol 1: 347–356.

Goad CC (1986): Precise positioning with the GPS. CERN Accelerator School, Applied Geodesy for Particle Accelerators, Geneva, Switzerland, April 14–18.

Goad CC, Goodman L (1974): A modified Hopfield tropospheric refraction correction model. Paper presented at the American Geophysical Union Annual Fall Meeting, San Francisco, California, December 12–17.

Gouldman MW, Herman BR, Swift ER (1986): Absolute station position solutions for sites involved in the spring 1985 GPS precision baseline test. In: Proceedings of the Fourth International Geodetic Symposium on Satellite Positioning, Austin, Texas, April 28 – May 2, vol 2: 1045–1057.

Gouldman MW, Herman BR, Weedon DL (1989): Evaluation of GPS production ephemeris and clock quality. In: Proceedings of the Fifth International Geodetic Symposium on Satellite Positioning, Las Cruces, New Mexico, March 13–17, vol 1: 210–222.

Grafarend EW, Schwarze V (1991): Relativistic GPS positioning. In: Caputo M, Sansò F (eds): Proceedings of the geodetic day in honor of Antonio Marussi. Accademia Nazionale dei Lincei, Rome. Atti dei Convegni Lincei, vol 91: 53–66.

Grant DB (1988): Combination of terrestrial and GPS data for earth deformation studies in New Zealand. University of New South Wales, School of Surveying, Kensington, Australia.

Gurtner W, Beutler G, Rothacher M (1989): Combination of GPS observations made with different receiver types. In: Proceedings of the Fifth International Geodetic Symposium on Satellite Positioning, Las Cruces, New Mexico, March 13–17, vol 1: 362–374.

Gurtner W, Mader G (1990): Receiver independent exchange format version 2. GPS Bulletin, 3(3): 1–8.

Gurtner W, Mader G, McArthur D (1989): A common exchange format for GPS data. GPS Bulletin, 2(3): 1–11.

Hartl P, Schöller W, Thiel K-H (1985): GPS related activities of the INS. In: Welsch WM, Lapine LA (eds): Proceedings of the Joint Meeting of FIG

Study Groups 5B and 5C on Inertial, Doppler and GPS Measurements for National and Engineering Survey, Munich, July 1-3. Schriftenreihe der Universität der Bundeswehr München, vol 20-2: 391–401.

Hatch R (1982): The synergism of GPS code and carrier measurements. In: Proceedings of the Third International Symposium on Satellite Doppler Positioning, New Mexico State University, February 8–12, vol 2: 1213–1231.

Hatch R (1986): Dynamic differential GPS at the centimeter level. In: Proceedings of the Fourth International Geodetic Symposium on Satellite Positioning, Austin, Texas, April 28 – May 2, vol 2: 1287–1298.

Hatch R (1990): Instantaneous ambiguity resolution. In: Schwarz KP, Lachapelle G (eds): Kinematic systems in geodesy, surveying, and remote sensing. Springer, New York Berlin Heidelberg London Paris Tokyo Hong Kong, 299–308 [Mueller II (ed): IAG Symposia Proceedings, vol 107].

Hatch R, Larson K (1985): MAGNET-4100 GPS survey program processing techniques and test results. In: Proceedings of the First International Symposium on Precise Positioning with the Global Positioning System, Rockville, Maryland, April 15–19, vol 1: 285–297.

Heckmann B (1985): Über die Auswirkung von relativistischen Effekten auf geodätische Messungen. Allgemeine Vermessungsnachrichten, 8–9: 329–336.

Hein GW (1990a): Bestimmung orthometrischer Höhen durch GPS und Schweredaten. Schriftenreihe der Universität der Bundeswehr München, vol 38-1: 291–300.

Hein GW (1990b): Kinematic differential GPS positioning: applications in airborne photogrammetry and gravimetry. In: Crosilla F, Mussio L (eds): Il sistema di posizionamento globale satellitare GPS. International Centre for Mechanical Sciences (CISM), Collana di Geodesia e Cartografia, Udine, Italy, 139–173.

Hein GW, Landau H, Baustert G (1988): Terrestrial and aircraft differential kinematic GPS positioning. In: Groten E, Strauß R (eds): GPS-techniques applied to geodesy and surveying. Springer, Berlin Heidelberg New York London Paris Tokyo, 307–348 [Bhattacharji S, Friedman GM, Neugebauer HJ, Seilacher A (eds): Lecture Notes in Earth Sciences, vol 19].

Hein GW, Leick A, Lambert S (1989): Integrated processing of GPS and gravity data. Journal of Surveying Engineering, 115(1): 15–33.

Heiskanen WA, Moritz H (1967): Physical Geodesy. Freeman, San Francisco London.

Henstridge F (1991): Getting started in GPS. Professional Surveyor, 11(4): 4–9.

Hilla SA (1986): Processing cycle slips in nondifferenced phase data from the Macrometer V-1000 receiver. In: Proceedings of the Fourth International Geodetic Symposium on Satellite Positioning, Austin, Texas, April 28 – May 2, vol 1: 647–661.

Hofmann-Wellenhof B (1990): Landesvermessung und Landinformation, Lecture Notes for Surveying. University of Technology, Graz, Austria.

Hofmann-Wellenhof B, Klostius W, Pesec P (1990): Real-time relative positioning with GPS. In: Proceedings of the Second International Symposium on Precise Positioning with the Global Positioning System, Ottawa, Canada, September 3–7, 1248–1256.

Hofmann-Wellenhof B, Lichtenegger H (eds) (1988): GPS – Von der Theorie zur Praxis. Mitteilungen der geodätischen Institute der Technischen Universität Graz, vol 62.

Hofmann-Wellenhof B, Remondi BW (1988): The antenna exchange: one aspect of high-precision GPS kinematic survey. In: Groten E, Strauß R (eds): GPS-techniques applied to geodesy and surveying. Springer, Berlin Heidelberg New York London Paris Tokyo, 261–277 [Bhattacharji S, Friedman GM, Neugebauer HJ, Seilacher A (eds): Lecture Notes in Earth Sciences, vol 19].

Holdridge DB (1967): An alternate expression for light time using general relativity. Jet Propulsion Laboratory, Supporting Research and Advanced Development, Space Programs Summary, 3: 37–48.

Hopfield HS (1969): Two-quartic tropospheric refractivity profile for correcting satellite data. Journal of Geophysical Research, 74(18): 4487–4499.

Hothem LD, Fronczek CJ (1983): Report on test and demonstration of Macrometer model V-1000 interferometric surveyor. FGCC Report: FGCC-IS-83-2.

Hwang PYC (1990): Kinematic GPS: resolving integer ambiguities on the fly. In: Proceedings of the IEEE Position Location and Navigation Symposium, Las Vegas, March 20–23, 579–586.

Jäger R, Mierlo J van (1991): Mathematische Modellbildungen bei der Integration von GPS-Konfigurationen in bestehende Netze. Deutscher Verein für Vermessungswesen, special issue: GPS und Integration von GPS in bestehende geodätische Netze, vol 38: 143–164.

Janes HW, Langley RB, Newby SP (1989): A comparison of several models for the prediction of tropospheric propagation delay. In: Proceedings of the Fifth International Geodetic Symposium on Satellite Positioning, Las Cruces, New Mexico, March 13–17, vol 2: 777–788.

Jones T (1989): NAVSTAR Global Positioning System – status and update. In: Proceedings of the Fifth International Geodetic Symposium on Satellite Positioning, Las Cruces, New Mexico, March 13–17, vol 1: 28–52.

Joos G (1956): Lehrbuch der Theoretischen Physik. Akademische Verlagsgesellschaft Geest & Portig K-G, Leipzig, 9th edition.

Jorgensen PS (1980): Combined pseudo range and Doppler positioning for the stationary NAVSTAR user. In: Proceedings of the Position Location and Navigation Symposium, IEEE Publication 80CH1597-4: 450–458.

Jorgensen PS (1989): An assessment of ionospheric effects on the GPS user. Navigation, 36(2): 195–204.

Kaniuth K (1986): A local model for estimating the tropospheric path delay at microwave frequencies. In: Proceedings of the Fourth International Geodetic Symposium on Satellite Positioning, Austin, Texas, April 28 – May 2, vol 1: 589–601.

Kaniuth K, Stuber K, Tremel H (1989): A comparative analysis of various procedures for modelling the tropospheric delay in a regional GPS network. In: Proceedings of the Fifth International Geodetic Symposium on Satellite Positioning, Las Cruces, New Mexico, March 13–17, vol 2: 767–776.

Kato T, Murata I, Tsuchiya A (1987): Effects of ionosphere on interferometric GPS observations. Earthquake Research Institute, University of Tokyo.

Kaula WM (1966): Theory of Satellite Geodesy. Blaisdell, Toronto.

Kinal GV, Singh JP (1990): An international geostationary overlay for GPS and GLONASS. Navigation, 37(1): 81–93.

King RW, Masters EG, Rizos C, Stolz A, Collins J (1987): Surveying with Global Positioning System. Dümmler, Bonn.

Klees R (1990): Anwendung des NAVSTAR/Global Positioning System in der kanadischen Landes-, Kataster- und Stadtvermessung. Allgemeine Vermessungsnachrichten, 97(3): 117–120, and 97(4): 138–157.

Kleusberg A (1990a): A review of kinematic and static GPS surveying procedures. In: Proceedings of the Second International Symposium on Precise Positioning with the Global Positioning System, Ottawa, Canada, September 3–7, 1102–1113.

Kleusberg A (1990b): Comparing GPS and GLONASS. GPS World, 1(6): 52.

Klobuchar J (1986): Design and characteristics of the GPS ionospheric time-delay algorithm for single-frequency users. In: Proceedings of the IEEE Position Location and Navigation Symposium, Las Vegas, November 4–7.

Knuth DE (1978): The art of computer programming – fundamental algorithms. Addison Wesley, Reading (Massachusetts) Menlo Park (California) London Amsterdam Don Mills (Ontario) Sydney, vol 1, 2nd edition.

Kozai Y (1959): On the effects of the sun and the moon upon the motion of a close earth satellite. Smithsonian Astrophysical Observatory, Special Report vol 22.

Kreyszig E (1968): Advanced engineering mathematics. Wiley, New York London Sydney, 2nd edition.

Lachapelle G (1990): GPS observables and error sources for kinematic positioning. In: Schwarz KP, Lachapelle G (eds): Kinematic systems in geodesy, surveying, and remote sensing. Springer, New York Berlin Heidelberg London Paris Tokyo Hong Kong, 17–26 [Mueller II (ed): IAG Symposia Proceedings, vol 107].

Lachapelle G (1991): GPS developments and impacts. Lecture Notes.

Lachapelle G, Beck N, Héroux P (1982): NAVSTAR/GPS single point positioning using pseudo-range and Doppler observations. In: Proceedings of the Third International Symposium on Satellite Doppler Positioning, New Mexico State University, February 8–12, vol 2: 1079–1091.

Lachapelle G, Hagglund J, Falkenberg W, Bellemare P, Casey M, Eaton M (1986): GPS land kinematic positioning experiments. In: Proceedings of the Fourth International Geodetic Symposium on Satellite Positioning, Austin, Texas, April 28 – May 2, vol 2: 1327–1344.

Landau H (1988): Zur Nutzung des Global Positioning Systems in Geodäsie und Geodynamik: Modellbildung, Software-Entwicklung und Analyse. Schriftenreihe der Universität der Bundeswehr München, vol 36.

Landau H (1990): GPS processing techniques in geodetic networks. In: Proceedings of the Second International Symposium on Precise Positioning with the Global Positioning System, Ottawa, Canada, September 3–7, 373–386.

Langley RB (1991): The GPS receiver: an introduction. GPS World, 2(1): 50–53.

Lanyi G (1984): Tropospheric calibration in radio interferometry. In: Proceedings of the International Symposium on Space Techniques for Geodynamics, Sopron, Hungary, July 9–13, vol 2: 184–195.

Leick A (1990): GPS satellite surveying. Wiley, New York Chichester Brisbane Toronto Singapore.

Lichten SM, Bertiger WI, Lindqwister UJ (1989): The effect of fiducial network strategy on high-accuracy GPS orbit and baseline determination. In: Proceedings of the Fifth International Geodetic Symposium on Satellite Positioning, Las Cruces, New Mexico, March 13–17, vol 1: 516–525.

Lichten SM, Neilan RE (1990): Global networks for GPS orbit determination. In: Proceedings of the Second International Symposium on Precise Positioning with the Global Positioning System, Ottawa, Canada, September 3–7, 164–178.

Lichtenegger H (1991): Über die Auswirkung von Koordinatenänderungen in der Referenzstation bei relativen Positionierungen mittels GPS. Österreichische Zeitschrift für Vermessungswesen und Photogrammetrie, 79(1): 49–52.

Lichtenegger H, Hofmann-Wellenhof B (1989): GPS-data preprocessing for cycle-slip detection. In: Bock Y, Leppard N (eds): Global Positioning System: an overview. Springer, New York Berlin Heidelberg London Paris Tokyo Hong Kong, 57–68 [Mueller II (ed): IAG Symposia Proceedings, vol 102].

Loomis P (1989): A kinematic GPS double-differencing algorithm. In: Proceedings of the Fifth International Geodetic Symposium on Satellite Positioning, Las Cruces, New Mexico, March 13–17, vol 2: 611–620.

McArthur D, Beck N, Lochhead K, Delikaraoglou D (1985): Precise relative positioning with the Macrometer V-1000 interferometric surveyor: experiences at the Geodetic Survey of Canada. In: Proceedings of the First International Symposium on Precise Positioning with the Global Positioning System, Rockville, Maryland, April 15–19, vol 2: 521–532.

McCarthy DD (ed) (1989): IERS Standards (1989). Observatoire de Paris, International Earth Rotation Service, Technical Note 3.

MacDoran PF, Miller RB, Buennagel LA, Whitcomb JH (1985): Codeless systems for positioning with NAVSTAR-GPS. In: Proceedings of the First International Symposium on Precise Positioning with the Global Positioning System, Rockville, Maryland, April 15–19, vol 1: 181–190.

Mader GL (1986): Dynamic positioning using GPS carrier phase measurements. Manuscripta geodaetica, 11: 272–277.

Mader GL (1990): Ambiguity function techniques for GPS phase initialization and kinematic solutions. In: Proceedings of the Second International Symposium on Precise Positioning with the Global Positioning System, Ottawa, Canada, September 3–7, 1233–1247.

Marini JW, Murray CW (1973): Correction of laser range tracking data for atmospheric refraction at elevations above 10 degrees. NASA/GSCF X-591-73-351.

Melbourne WG (1985): The case for ranging in GPS-based geodetic systems. In: Proceedings of the First International Symposium on Precise Positioning with the Global Positioning System, Rockville, Maryland, April 15–19, vol 1: 373–386.

Melchior P (1978): The tides of the planet earth. Pergamon Press, Oxford New York Toronto Sydney Paris Frankfurt.

Merminod B, Grant DB, Rizos C (1990): Planning GPS-surveys using appropriate precision indicators. CISM Journal, 44(3): 233–24.

Meyerhoff SL, Evans AG (1986): Demonstration of the combined use of GPS pseudorange and Doppler measurements for improved dynamic positioning. In: Proceedings of the Fourth International Geodetic Symposium on Satellite Positioning, Austin, Texas, April 28 – May 2, vol 2: 1397–1409.

Milliken RJ, Zoller CJ (1980): Principle of operation of NAVSTAR and system characteristics. In: The Institute of Navigation: Global Positioning System, vol 1: 3–14.

Minkel DH (1989): Demonstration and discussion of the pseudo-kinematic method. In: Proceedings of the Fifth International Geodetic Symposium on Satellite Positioning, Las Cruces, New Mexico, March 13–17, vol 2: 577–588.

Montenbruck O (1984): Grundlagen der Ephemeridenrechnung. Sterne und Weltraum Vehrenberg, München.

Montgomery H (1991): GPS – the next generation. GPS World, 2(10): 12–16.

Moritz H (1977): Introduction to interpolation and approximation. Lecture Notes of the Second International Summer School in the Mountains, Ramsau, Austria, August 23 – September 2.

Moritz H (1980): Advanced Physical Geodesy. Wichmann, Karlsruhe.

Moritz H, Mueller II (1988): Earth rotation – theory and observation. Ungar, New York.

Mueller II (1991): International GPS Geodynamics Service. GPS Bulletin, 4(1): 7–16.

Mueller II, Archinal B (1981): Geodesy and the Global Positioning System. Paper presented at the IAG Symposium on Geodetic Networks and Computations, Munich, August 31 – September 5.

Munck JC de (1989): Kalman filtering. In: Bakker G, Munck JC de, Strang van Hees GL: Radio positioning at sea. Delft University.

Nautical Almanac Office (1983): The astronomical almanac for the year 1984. US Government Printing Office, Washington.

Nieuwejaar PW (1988): GPS signal structure. In: The Navstar GPS system. US Department of Commerce, National Technical Information Service, 5.1–6.

Oswald J, Mitchell J, Whiting L (1986): Simple differential techniques using the Trimble 4000A GPS locator. In: Proceedings of the Fourth International Geodetic Symposium on Satellite Positioning, Austin, Texas, April 28 – May 2, vol 1: 503–511.

Payne CR (1982): NAVSTAR Global Positioning System: 1982. In: Proceedings of the Third International Symposium on Satellite Doppler Positioning, New Mexico State University, February 8–12, vol 2: 993–1021.

Perreault PD (1980): Description of the Global Positioning System (GPS) and the STI receivers. CSTG Bulletin vol 2, Technology and Mission Developments.

Pesec P, Stangl G (1990): TI-4100 PROM-type automation for unattended operation in international GPS orbit-service programs. In: Colic K, Pesec P, Sünkel H (eds) (1990): Proceedings of the First International Symposium on Gravity Field Determination and GPS-Positioning in the Alps-Adria Area. Mitteilungen der Geodätischen Institute der Technischen Universität Graz, vol 67: 336–344.

Prescott W, Davis J, Svarc J (1989): Height of L2 phase center for TI-antennas. GPS Bulletin, 2(2): 13.

Prilepin MT (1989): Improvement of accuracy of single-point determinations. In: Proceedings of the Fifth International Geodetic Symposium on Satellite Positioning, Las Cruces, New Mexico, March 13–17, vol 1: 462–473.

Rahnemoon M (1988): Ein Korrekturmodell für Mikrowellenmessungen zu Satelliten. German Geodetic Commission, Munich, series C, vol 335.

Reichert G (1986): A new water-vapor radiometer design. In: Proceedings of the Fourth International Geodetic Symposium on Satellite Positioning, Austin, Texas, April 28 – May 2, vol 2: 603–613.

Remondi BW (1984): Using the Global Positioning System (GPS) phase observable for relative geodesy: modeling, processing, and results. University of Texas at Austin, Center for Space Research.

Remondi BW (1985): Global Positioning System carrier phase: description and use. Bulletin Géodésique, 59: 361–377.

Remondi BW (1986): Performing centimeter-level surveys in seconds with GPS carrier phase: initial results. In: Proceedings of the Fourth International Geodetic Symposium on Satellite Positioning, Austin, Texas, April 28 – May 2, vol 2: 1229–1249.

Remondi BW (1988): Kinematic and pseudo-kinematic GPS. In: Proceedings of the Satellite Division Conference of the Institute of Navigation, Colorado Springs, Colorado, September 21–23.

Remondi BW (1989): Extending the National Geodetic Survey standard GPS orbit formats. National Information Center, Rockville, Maryland, NOAA Technical Report NOS 133, NGS 46.

Remondi BW (1990a): Pseudo-kinematic GPS results using the ambiguity function method. National Information Center, Rockville, Maryland, NOAA Technical Memorandum NOS NGS-52.

Remondi BW (1990b): Recent advances in pseudo-kinematic GPS. In: Proceedings of the Second International Symposium on Precise Positioning with the Global Positioning System, Ottawa, Canada, September 3–7, 1114–1137.

Remondi BW (1991a): Kinematic GPS results without static initialization. National Information Center, Rockville, Maryland, NOOA Technical Memorandum NOS NGS-55.

Remondi BW (1991b): NGS second generation ASCII and binary orbit formats and associated interpolation studies. Paper presented at XX General Assembly of IUGG, Vienna, August 11-24.

Remondi BW (1991c): The Global Positioning System. The Military Engineer, 84(545): 31–36.

Remondi BW, Hofmann-Wellenhof B (1989a): Accuracy of Global Positioning System broadcast orbits for relative surveys. National Information Center, Rockville, Maryland, NOAA Technical Report NOS 132, NGS 45.

Remondi BW, Hofmann-Wellenhof B (1989b): GPS broadcast orbits versus precise orbits: a comparison study. GPS Bulletin, 2(6): 8–13.

Richardus P, Adler RK (1972): Map projections for geodesists, cartographers and geographers. North-Holland, Amsterdam London.

Rizos C, Stolz A (1985): Force modelling for GPS satellite orbits. In: Proceedings of the First International Symposium on Precise Positioning with the Global Positioning System, Rockville, Maryland, April 15–19, vol 1: 87–96.

Rizos C, Govind R, Stolz A, Luck JMcK (1987): The Australian GPS orbit determination pilot project. Australian Journal for Geodesy, Photogrammetry and Surveying, vol 46 & 47: 17–40.

Rocken C, Meertens CM (1989): GPS antenna and receiver tests: multipath reduction and mixed receiver baselines. In: Proceedings of the Fifth International Geodetic Symposium on Satellite Positioning, Las Cruces, New Mexico, March 13–17, vol 1: 375–385.

Rockwell International Corporation (1984): Navstar GPS space segment. Downey, California, Interface Control Document ICD-GPS-200.

Rogers AEE, Knight CA, Hinteregger HF, Whitney AR, Counselman CC, Shapiro II, Gourevitch SA, Clark TA (1978): Geodesy by radio interferometry. Determination of a 1.24 km baseline vector with 5 millimeter repeatability. Journal of Geophysical Research, 83: 325–333.

Rothacher M, Beutler G, Gurtner W, Schildknecht T, Wild U (1989): Results of the 1984, 1986, and 1988 Alaska GPS campaigns. In: Proceedings of the Fifth International Geodetic Symposium on Satellite Positioning, Las Cruces, New Mexico, March 13–17, vol 1: 554–566.

Rutscheidt EH, Roth BD (1982): The NAVSTAR Global Positioning System. CSTG Bulletin vol 4, International Activities including Proceedings of Symposium 4e, IAG General Meeting Tokyo.

Saastamoinen II (1973): Contribution to the theory of atmospheric refraction. Bulletin Géodésique, 107: 13–34.

Scherrer R (1985): The WM GPS primer. WM Satellite Survey Company, Wild, Heerbrugg, Switzerland.

Schmitt G, Illner M, Jäger R (1991): Transformationsprobleme. Deutscher Verein für Vermessungswesen, special issue: GPS und Integration von GPS in bestehende geodätische Netze, vol 38: 125–142.

Schödlbauer A, Krack K, Glasmacher H (1989): Densification of horizontal networks by GPS. In: Proceedings of the Fifth International Geodetic Symposium on Satellite Positioning, Las Cruces, New Mexico, March 13–17, vol 2: 1090–1103.

308

Schupler BR, Clark TA (1991): How different antennas affect the GPS observable. GPS World, 2(10): 32–36.

Schwarz KP (1983): Inertial surveying and geodesy. Reviews of Geophysics and Space Physics, 21(4): 878–890.

Schwarz KP (1987): Geoid profiles from an integration of GPS satellite and inertial data. Bollettino di geodesia e scienze affini, 2: 117–131.

Schwarz KP, Arden DAG (1985): A comparison of adjustment and smoothing methods for inertial networks. In: Schwarz KP (ed) (1985): Proceedings of the Symposium on Inertial Technology for Surveying and Geodesy, Banff, Canada, September 16–20, vol 1: 257–271.

Schwarz KP, Cannon ME, Wong RVC (1987): The use of GPS in exploration geophysics – a comparison of kinematic models. Paper presented at the XIX General Assembly of the IUGG at Vancouver, Canada, August 10–22.

Schwarz KP, Lachapelle G (eds) (1990): Kinematic systems in geodesy, surveying, and remote sensing. Springer, New York Berlin Heidelberg London Paris Tokyo Hong Kong, 299–308 [Mueller II (ed): IAG Symposia Proceedings, vol 107].

Schwiderski EW (1981): Global ocean tides, atlas of tidal charts and maps. Naval Surface Weapons Center (NSWC), Virginia.

Seeber G (1989): Satellitengeodäsie: Grundlagen, Methoden und Anwendungen. Walter de Gruyter, Berlin New York.

Sims ML (1985): Phase center variation in the geodetic TI4100 GPS receiver system's conical spiral antenna. In: Proceedings of the First International Symposium on Precise Positioning with the Global Positioning System, Rockville, Maryland, April 15–19, vol 1: 227–244.

Smith ID (1964): Satellite hyperbolic navigating system. U.S. Patent Office, Patent no. 3,126,545.

Snay RA (1986): Network design strategies applicable to GPS surveys using three or four receivers. Bulletin Géodésique, 60(1): 37–50.

Spilker JJ (1980): GPS signal structure and performance characteristics. In: The Institute of Navigation: Global Positioning System, vol 1: 29–54.

Stangl G, Hofmann-Wellenhof B, Pesec P, Sünkel H (1991): Austrian GPS reference network – concept, realization, and first results. Paper presented at XX General Assembly of IUGG, Vienna, August 11–24.

Stein WL (1986): NAVSTAR Global Positioning System 1986 status and plans. In: Proceedings of the Fourth International Geodetic Symposium on Satellite Positioning, Austin, Texas, April 28 – May 2, vol 1: 37–49.

Strange WE (1985): High-precision, three-dimensional differential positioning using GPS. In: Proceedings of the First International Symposium on Precise Positioning with the Global Positioning System, Rockville, Maryland, April 15–19, vol 2: 543–548.

Swift ER (1985): NSWC'S GPS orbit/clock determination system. In: Proceedings of the First International Symposium on Precise Positioning with the Global Positioning System, Rockville, Maryland, April 15–19, vol 1: 51–62.

Tallqvist S (1985): The GPS microstrip antenna properties; reduction of multipath contamination and other interference by an RF absorbent ground plane. In: Proceedings of the Second SATRAPE Meeting, Saint-Mandé, France, November 4–6, 46–59.

Tranquilla JM (1986): Multipath and imaging problems in GPS receiver antennas. In: Proceedings of the Fourth International Geodetic Symposium on Satellite Positioning, Austin, Texas, April 28 – May 2, vol 1: 557–571.

Unguendoli M (1990): A rational approach to the use of a large number of GPS receivers. Bulletin Géodésique, 64(4): 303-312.

Walser F (1988): Ionosphäreneinfluss bei GPS-Messungen. Eidgenössische Technische Hochschule Zürich, Institute of Geodesy and Photogrammetry, vol 147.

Wells D (1985): Recommended GPS terminology. In: Welsch WM, Lapine LA (eds): Proceedings of the Joint Meeting of FIG Study Groups 5B and 5C on Inertial, Doppler and GPS Measurements for National and Engineering Survey, Munich, July 1–3. Schriftenreihe der Universität der Bundeswehr München, vol 20-1: 179–207.

Wells DE, Beck N, Delikaraoglou D, Kleusberg A, Krakiwsky EJ, Lachapelle G, Langley RB, Nakiboglu M, Schwarz KP, Tranquilla JM, Vanicek P (1987): Guide to GPS positioning. Canadian GPS Associates, Fredericton, New Brunswick, Canada.

Wells D, Lachapelle G (1981): Impact of NAVSTAR/GPS on land and off-shore positioning in the 1980s. Paper presented at Canadian Petroleum Association Colloquium III, Petroleum Mapping and Surveys in the '80s, Banff, Alberta, October 14–16.

Westrop J, Napier ME, Ashkenazi V (1989): Cycle slips on the move: detection and elimination. In: Proceedings of ION GPS-89, The Second International Technical Meeting of the Satellite Division of The Institute of Navigation, Colorado Springs, Colorado, September 27–29, 31–34.

Wild U, Beutler G, Gurtner W, Rothacher M (1989): Estimating the ionosphere using one or more dual frequency GPS receivers. In: Proceedings of the Fifth International Geodetic Symposium on Satellite Positioning, Las Cruces, New Mexico, March 13–17, vol 2: 724–736.

Willis P, Boucher C (1990): High precision kinematic positioning using GPS at the IGN: recent results. In: Bock Y, Leppard N (eds): Global Positioning System: an overview. Springer, New York Berlin Heidelberg London Paris Tokyo Hong Kong, 340–350 [Mueller II (ed): IAG Symposia Proceedings, vol 102].

Wooden WH (1985): Navstar Global Positioning System: 1985. In: Proceedings of the First International Symposium on Precise Positioning with the Global Positioning System, Rockville, Maryland, April 15–19, vol 1: 23–32.

Wübbena G (1985): Software developments for geodetic positioning with GPS using TI 4100 code and carrier measurements. In: Proceedings of the First International Symposium on Precise Positioning with the Global Positioning System, Rockville, Maryland, April 15–19, vol 1: 403–412.

Wübbena G (1988): GPS carrier phases and clock modeling. In: Groten E, Strauß R (eds): GPS-techniques applied to geodesy and surveying. Springer, Berlin Heidelberg New York London Paris Tokyo, 381–392 [Bhattacharji S, Friedman GM, Neugebauer HJ, Seilacher A (eds): Lecture Notes in Earth Sciences, vol 19].

Wübbena G (1989): The GPS adjustment software package GEONAP – concepts and models. In: Proceedings of the Fifth International

Geodetic Symposium on Satellite Positioning, Las Cruces, New Mexico, March 13–17, vol 2: 452–461.

Wunderlich T (1992): Die gefährlichen Örter der Pseudostreckenortung. Technical University Hannover.

Yionoulis S M (1970): Algorithm to compute tropospheric refraction effects on range measurements. Journal of Geophysical Research, 75(36): 7636–7637.

Yiu KP, Crawford R, Eschenbach R (1984): A low-cost GPS receiver for land navigation. In: The Institute of Navigation: Global Positioning System, vol 2: 44–60.

Young LE, Neilan RE, Bletzacker FR (1985): GPS satellite multipath: an experimental investigation. In: Proceedings of the First International Symposium on Precise Positioning with the Global Positioning System, Rockville, Maryland, April 15–19, vol 1: 423–432.

Zhu SY, Groten E (1988): Relativistic effects in GPS. In: Groten E, Strauß R (eds): GPS-techniques applied to geodesy and surveying. Springer, Berlin Heidelberg New York London Paris Tokyo, 41–46 [Bhattacharji S, Friedman GM, Neugebauer HJ, Seilacher A (eds): Lecture Notes in Earth Sciences, vol 19].

Subject index